The Clinical
Psychology of Aging

The Clinical
Psychology of Aging

Edited by
MARTHA STORANDT
Washington University
St. Louis, Missouri

ILENE C. SIEGLER
Duke University Medical School
Durham, North Carolina
and
MERRILL F. ELIAS
University of Maine at Orono
Orono, Maine

PLENUM PRESS · **NEW YORK AND LONDON**

Library of Congress Cataloging in Publication Data

Main entry under title:

The clinical psychology of aging.

Includes bibliographical references and index.
   1. Geriatric psychiatry. 2. Psychological tests. 3. Psychotherapy. 4.
Clinical psychology. I. Storandt, Martha. II. Siegler, Ilene C. III. Elias,
Merrill F., 1938-   [DNLM: 1. Psychology, clinical—In old age.
WT150 C641]
RC451.4.A5C54   618.9'76'89   78-4566
ISBN 0-306-40001-4

© 1978 Plenum Press, New York
A Division of Plenum Publishing Corporation
227 West 17th Street, New York, N.Y. 10011

Printed in the United States of America

Preface

It has been estimated that there are at least 2,500,000 adults, 10% of the population above age 65, who are currently in need of some sort of mental health services (Kramer, Taube, and Redick, 1973). Other estimates are even higher (e.g., Pfeiffer, 1977). It is expected that this number will increase as the number of older adults increases over the next 40 years.

Probably less than 400 clinical psychologists are now providing services to this age group. The number of elderly patients actually seen by these psychologists is very, very small. One national survey found that of 353 psychologists who reported that they had older clients, only 495 individuals were seen for psychological testing and 1423 for psychotherapy in the one month just prior to the response (Dye, in press). Assuming that the same individuals were not seen for both testing and therapy within the one month period--a questionable assumption--approximately .08% of the at least two-and-one-half million older adults in need of psychological services are now being supplied with these services in some form or another. Thus, the need for increased involvement of clinical psychology with the aged is undeniable. However, few resources currently exist which will serve to increase the number of clinical psychologists trained to meet this need. Probably less than 100 clinical psychologists living today have received any kind of formal graduate training in the clinical psychology of the aging (Storandt, 1977). Training programs in the clinical psychology of aging now exist at only one or two universities. Internship facilities which will allow the student clinician to gain experience with older adults under competent supervision are almost nonexistent (Siegler, Gentry, and Edwards, in press). Post-graduate continuing education opportunities in this area are rare. Few professional journals read by clinicians carry articles on the clinical psychology of aging (Lawton, 1970).

A recent report to the Secretary of the Department of Health, Education and Welfare on manpower and training needs with respect to the mental health needs of older adults recommended a concerted

effort to produce 2,000 new clinical geropsychologists within the
next 10 years (Birren and Sloane, 1977). This would represent
greater than a five-fold increase in available manpower. This
still is not nearly enough psychologists in this speciality area
but certainly it represents a beginning. Legislation is now before the Congress to provide for independent reimbursement of
psychological services rendered to older adults through the Medi-
care Act. Such a provision will open the door to greater and
greater involvement of clinical psychologists in private practice
with older clients; however, these clinicians must be trained to
deal with the special needs and services required by elderly persons.

The training needs and efforts described above and the
broader efforts of the profession of psychology as a whole to deal
with the mental health needs of older adults require knowledge
which can be transmitted to the student and to the practicing
clinician. This book represents an effort to gather the current
literature concerning clinical geropsychology into one volume for
use in such educational programs. It is hoped that it will serve
as a resource to instructors, as a text to students, and as a
reference to practicing clinicians. Other disciplines concerned
with the mental health of older adults such as psychiatry, nursing,
and social work may also find it useful.

It also is hoped that this book will serve to focus research
efforts in the clinical psychology of aging. As will become clear
to the reader who progresses through the various chapters, our
knowledge of the psychopathology found in older adults is spotty.
We are faced with difficult practical and theoretical issues in
terms of differential diagnosis. Current knowledge of effective
therapeutic techniques applicable to older adults faced with
psychological problems is limited. Two concerns common to all
research on aging are the reseacher's definition of aging and the
selection of appropriate control groups. These issues are
touched on by the authors who contributed to this volume and may
account for some of the discrepancies in the literature reviewed
within and between the various chapters. It will become clear
to the reader that these issues are complex, cannot be dealt with
easily, and require concerted and extended research efforts. It
is as if psychology has put off dealing with clinical geropsychology because the profession recognizes that it will be sorely taxed
to find answers to many of the questions which now exist in this
subspeciality.

Many of the issues which must be resolved in this new sub-
discipline may require multi- and interdisciplinary approaches.
For example, effective treatment of senile dementia may not be
possible until geneticists, physiologists, biochemists, and neuro-

pathologists provide us with a better understanding of the basic
process underlying this problem. Successful solutions to the
psychological problems which may be associated with retirement may
hinge on wide social changes which may be dealt with by sociolo-
gists and other social scientists. However, the clinical psychologist, with a foundation in both biobehavioral and social sciences, may act as a translator and thus serve as a catalyst in
these research undertakings.

The emphasis of most of the chapters in this book is upon a
review of the research literature concerning the various content
areas and an attempt to delineate areas in which more research
effort is required. Indeed, it is assumed that the reader has a
thorough grounding in research methodology as well as in psychological principles and theories. This volume is not a "how-to-do-
it" handbook for practicing clinicians. Clearly, it is not a
guide to universal happiness and mental well being in old age for
the layman.

With respect to the organization of the book, it is divided
into three sections. The first deals with assessment issues as
they relate to cognitive functioning in older adults. Research
psychologists interested in gerontology in the past have been experimental psychologists primarily and have focused on age-related
changes in cognitive functions such as intelligence, perception,
learning, memory, and attention. However, much of this research
has been addressed to the question of how these processes change
as a function of age in the "normal" individual. Less emphasis
has been placed upon changes in cognitive function as a result of
pathology and especially organic pathology. Thus, it would seem
that a primary focus of future research may be upon brain functions and their influence on social and interpersonal relationships
as well as upon the ability of the older adult to process environmental stimuli.

The second section of this book deals with personality assessment in older adults. Research on personality in old age has been
hampered in the past by the difficulty of measuring such a complex
concept as personality (this problem faces all of psychology)
and by the lack of a body of theory related to personality in
later life. However, the personality which the older adult brings
to the clinical setting certainly influences the type of treatment chosen and its success. There may be certain personality
traits or characteristics which are felt to be typical of older
adults in general but which are indicative of pathology in the
individual. For example, the older adult is often described as
rigid and/or cautious. However, rigidity or cautiousness may not
be a personality trait characteristic of all older adults. In
some older persons it may be a defensive mechanism which the

individual uses to order an environment which has become increasingly difficult to deal with because of changes occurring as a
function of progressive brain dysfunction. In many it may reflect
an adaptive response to changing environments and social situations.
Thus, it is important to know what kinds of personality changes,
if any, occur with increasing age, how certain personality characteristics interact with various types of psychopathology, and
what types of behavior may be considered pathological.

The final section of this volume is devoted to a review of
the literature concerning research on the application of various
types of therapeutic procedures to older adults. Psychodynamic,
organic, and behavioristic techniques are covered in separate
chapters, while a potpourri of procedures are described in the
final paper. Therapy, of course, is the ultimate goal of clinical
psychology and the perceived difficulty of treating patients with
deteriorative nervous system disorders, which in reality occur in
only a minority of the elderly, may be the penultimate cause for
the short shrift clinical psychology has given the aged in the
past. Only 5% of all older adults are institutionalized and many
older adults living in the community live out their life span
without ever demonstrating any clinical signs of organic brain
dysfunction. Thus, it is foolish to ignore the needs of the large
segment of older adults who have treatable and relatively simple
problems in living which could be effectively dealt with by clinical psychology at this moment in time. Further, unless we _try_ to
treat the patient who suffers from specific and well localized
brain damage and/or diffuse brain impairment, we will never learn
how to do so.

The editors would like to express their appreciation to those
who helped make this book possible. These include Bruce Layton,
who read much of the text and supplied many helpful and useful
comments; Nancy Miller, whose patience and skill in preparing
the typescript was incomparable; and Wanda Meek, who always knew
where everything was. Finally, our thanks to the authors of the
chapters contained in this volume. Their dedication to gerontology and concern about the well-being of older adults will make
possible a true science of clinical gerontology.

January 16, 1978   Martha Storandt
St. Louis, Missouri   Ilene C. Siegler
                                                    Merrill F. Elias

REFERENCES

Birren, J. E., and Sloane, R. B. Manpower and training needs in
    mental health and illness of the aging. A report to the
    Gerontological Society for the Committee to Study Mental
    Health and Illness of the Elderly of the Secretary of the
    Department of Health Education and Welfare. Los Angeles:
    Ethel Percy Andrus Gerontology Center, University of South-
    ern California, 1977.
Dye, C. J. Psychologists' role in the provision of mental health
    care for the elderly. Professional Psychologist, in press.
Kramer, M., Taube, C. A., and Redick, R. W. Patterns of use of
    psychiatric facilities by the aged: Past, present and future.
    In C. Eisdorfer and M. P. Lawton (Eds.), The psychology of
    adult development and aging. Washington, D.C.: American
    Psychological Association, 1973.
Lawton, M. P. Gerontology in clinical psychology, and vice-versa.
    Aging and Human Development, 1970, 1, 147-159.
Pfeiffer, E. Psychopathology and social pathology. In J. E.
    Birren and K. W. Schaie (Eds.), Handbook of the psychology
    of aging. New York: Van Nostrand Reinhold, 1977.
Siegler, I. C., Gentry, W. D., and Edwards, C. D. Training in
    geropsychology: A survey of graduate and internship train-
    ing programs. Professional Psychology, in press.
Storandt, M. Graduate education in gerontological psychology:
    Results of a survey. Educational Gerontology, 1977, 2,
    141-146.

Contents

SECTION 3

THERAPY WITH THE AGED

Section 1
Cognitive Assessment

INTRODUCTION TO SECTION 1:   COGNITIVE ASSESSMENT

Merrill F. Elias

University of Maine at Orono

The chapters in this section deal with cognitive assessment. Eisdorfer and Cohen's chapter on "The Cognitively Impaired Elderly" sets the stage for the chapters which follow. It reviews the disorders of functional and organic origin that result in impairment of cognitive functioning in the elderly. This is particularly important information for the clinical psychologist who wishes to become more sensitive to the range of cognitive deficits which occur in "unhealthy aging individuals" as opposed to "healthy aging individuals." Variants of Alzheimer's disease and cerebrovascular disease are discussed as are other forms of disease affecting cognitive functioning either directly or indirectly.

Eisdorfer and Cohen make important distinctions among irreversible and reversible behavioral consequences of disease and discuss traditional problems of differential diagnosis in the context of elderly populations. There is, for example, an excellent discussion of the problems involved in the differentiating of effects of depression from those related to brain impairment per se. Eisdorfer and Cohen are prepared via disciplinary background, research, and training to facilitate an integration of the medical and nonmedical frame of reference in an approach to differential diagnosis that considers biological, social, genetic, and psychological factors. For example, there is a discussion of the medical examination and clinical laboratory tests in relationship to formalized cognitive evaluation techniques and interview.

The issue of formalized cognitive evaluation is addressed more specifically in the chapters on "Assessment of Altered Brain Function in the Elderly" (Kahn and Miller) and in "Neuropsychologi-

cal Evaluation in Older Persons" (Klisz). Neither the Kahn-
Miller chapter nor the Klisz chapter could have been as effective
in isolation as they are together. These chapters identify needed
areas of research in the assessment of brain damage and discuss
specific tests and test batteries which may be useful. In addition, they point out deficiencies in certain tests and test batteries under specific situations and for certain populations of
elderly persons. What is particularly interesting about these two
chapters is that they reflect divergent and yet unstated differences in philosophy of testing which are, in part, rooted in the
history of the development of psychology and psychiatry, the historical argument with regard to mass action versus specific localization of impaired brain function, and the traditional emphasis on
the development and refinement of single tests of brain damage versus a more contemporary emphasis on test batteries involving many
samples of different kinds of behaviors (the neuropsychological
approach).

It will be obvious to the reader that the Kahn-Miller chapter
is written more from the perspective of psychiatry than is the
Klisz chapter. There is more emphasis on a limited number of tests
dealing with commonalities of brain disorder in the former chapter
and more emphasis on the test battery approach in the latter.
However, both chapters reflect a sensitivity to the advantages
and limitations of the approaches they have chosen to emphasize
and document these positive and negative features with references
to the clinical literature. The Kahn-Miller chapter seems to
speak from the experience of clinical practice more than does the
Klisz chapter. The Klisz chapter seems more closely related to
basic research on cognitive functioning in elderly persons. This
difference in emphasis reflects not only the experience of the
authors but the state of the art of the two approaches illustrated
by these chapters. Hopefully, the reader will be sympathetic to
the practical clinical problems related to the use of the Reitan
Neuropsychological test battery with aged persons suffering from
multiple functional and/or organically-related factors, but will
at the same time be cognizant of the usefulness of the test battery
approach for the localizing of specific lesions.

Kahn and Miller call attention to the fact that subjective
analysis of the patient's answers to the Mental Status Questionnaire can be useful in terms of differentiating between chronic
and irreversible brain damage. Similarly, they emphasize the
efficiency with which simple tests may be given to some older persons. On the other hand, it is clear that efficiency is not the
only criteria for the development of maximally useful test batteries. Thus it would appear that development of a modified
neuropsychological test battery suitable for older adults is not
an unreasonable objective if the goal of diagnosis is to identify

specific functional areas of brain impairment and to make inferences with regard to locus, size, and type of lesion. The clinician who would undertake this task should find the chapters in
this section most useful in terms of relevant issues and timely
research.

G. W. Wood's chapter on the elderly alcoholic could have been
placed in either the section on treatment or personality assessment as he speaks to needed research in these areas as they pertain to alcoholism. The chapter was ultimately included in this
section because the chapter deals with the adverse effects of prolonged alcohol consumption on cognitive processes and emphasizes
the need to identify elderly and middle aged persons who function
poorly on tests of cognitive ability and brain integrity due to
excess consumption of alcohol rather than specific lesions or
deteriorating nervous system diseases. Wood points out that the
elderly alcoholic is not necessarily a young alcoholic who has
aged and argues convincingly that a decline in the rate of alcoholism with advancing age does not indicate that alcohol abuse in
the elderly is not a serious problem.

The effects of alcohol on behavior, problems of differential
diagnosis, and recent research developments with regard to understanding the relationship between reversible and irreversible symptoms of organic brain damage are discussed. These topics are particularly relevant to Eisdorfer and Cohen's chapter on differential diagnosis.

In general, the papers in this section emphasize the fact that
psychiatry and psychology have important contributions to make in
the area of differential diagnosis and cognitive assessment, and
that psychiatrists and psychologists have much to learn from each
other with respect to the best way to accomplish this task. Hopefully, this book will represent one of many steps in this direction.

THE COGNITIVELY IMPAIRED ELDERLY: DIFFERENTIAL DIAGNOSIS

Carl Eisdorfer

Department of Psychiatry and Behavioral Sciences
University of Washington

and

Donna Cohen

Department of Psychiatry and Behavioral Sciences
University of Washington

The dramatic changes in the age structure of the U.S. population and the rest of the world over the past 150 years have profound significance for mankind. The aged, who have become one of the fastest growing components of the population, are at an increasing risk for medical and psychiatric diseases. Comprising as they do 10% of the population, the aged account for 30% of the health care costs in the United States. They account for 14% of the outpatient visits to health care facilities and they are admitted to general hospitals 2.5 times the rate of young adults. Older persons occupy 90% of the approximately 1.2 million long term care facility beds, and there are estimates of millions more in the community who could benefit from appropriate care.

Unfortunately, dealing with the elderly who are at increased risk for psychiatric disorders (Kramer, Taube, and Redick, 1973) has not been popular with health professionals who have tended to be nihilistic, regarding the aged as hopeless and helpless and unworthy of careful evaluation and treatment. This attitude, in conjunction with the reluctance of older persons to identify problems as psychiatric disorders, makes the assessment of the true prevalence of psychiatric diseases in the elderly difficult. The prevalence of psychiatric disorders in the elderly living in the community is reported to range from 20-45% (Bremer, 1951;

Busse, Dovenmuhle, and Brown, 1960; Busse and Pfeiffer, 1975; Helgason, 1971; Kay, Beamish, and Roth, 1964).

Psychiatric disorders may be associated with a particularly
high rate of chronic and acute physical illnesses which may produce crippling disability and hospitalization (Lowenthal and Berkman, 1967). However, the utilization of psychiatric outpatient
facilities by the aged is far less than that of the young (Redick, Kramer,
and Taube, 1973). Estimates are that as many as 50% of elderly medical and surgical patients show some psychopathology (Schuckit,
Miller, and Hahlbaum, 1974; Busse and Pfeiffer, 1975). Schuckit
(1977b) has reported that 22% of 327 older persons admitted to an
acute medical or surgical ward of a VA Hospital over a one year
period met the criteria for a psychiatric disorder. The rate was
even higher (30%) among the subgroup with cardiac problems.

Geriatric alcoholism and drug abuse are often overlooked.
Although the true prevalence of alcoholism in the elderly is unknown, about 20% of older medical inpatients and 10-50% of medical
outpatients have problems related to alcohol (Schuckit, 1977a).
With regard to opiate abuse it has been estimated that 5% of
individuals in methadone maintenance are 45 years and older (Capel
and Stewart, 1971). Knowing or inadvertant abuse of drugs, other
than the opiates, is a major problem. Persons age 55 and older
are the largest consumers of drugs (Pascarelli and Fisher, 1974).
However, older patients may not take drugs as prescribed and consequently may suffer a drug reaction. Furthermore, there is an
increased likelihood of adverse reactions in the elderly to drugs
in normal dosages (Raskind and Eisdorfer, 1976a; Morrant, 1975;
Cooper, 1975). It is also not uncommon for older persons living
close together in housing projects to share drugs and collect
drugs for use long after the expiration dates (Raskind and Eisdorfer, 1976a; Subby, 1975).

Some of the drugs often abused by the elderly are barbituates
(Subby, 1975), laxatives (Cummings, Slader, and James, 1974),
and aspirin (Morrant, 1975). These drugs and a variety of others,
if abused, can lead to confusion and disorientation. The combination of prescribed medication, over-the-counter drugs, and alcohol
may also lead to drug intoxication, confusion, and the appearance
of cognitive dysfunction.

Frequently, psychiatric disorders are overlooked and are not
recorded and it is estimated that there may be at least a 30%
rate of incorrect diagnosis and treatment of the elderly which
would result in even greater morbidity than that documented (Kidd,
1962). Mental changes may be noted but they are regarded as part
of the aging process and are seen as irreversible. As a consequence, differential diagnosis and treatment for these disorders

which might alleviate the impairment or disability is either overlooked or disregarded.

Extension of knowledge regarding the nature and course of
psychopathology in the elderly and the development of new forms of
therapeutic intervention offer the greatest hope for prolonging
the productive life of the increasingly larger aged proportion of
the population who are at risk for significant mental impairment.
This chapter discusses the current approach to differential diagnosis of the cognitively and emotionally impaired elderly, and
focuses primarily upon the brain syndromes, depression, and paranoid behavior. In addition, relevant clinical, epidemiological,
behavioral, psychological, and biochemical studies contributing
to our knowledge of diagnosis and treatment will be presented.

DISORDERS OF COGNITIVE FUNCTIONING

In an effort to avoid the confusion of terms such as senile
dementia, organic brain syndrome and senile psychosis, we will
refer to a variety of conditions to be known collectively as cognitive disorders. The most salient feature of these disorders,
despite their clinically observed heterogeneity, is progressive impairment of cognition, i.e., attention, learning and memory. Additional symptoms which present frequently, but not always, include
hallucinations, delusions, aphasias, difficulty in abstraction, emotional lability, and depression. In specific instances other
neurologic, affective, or behavioral signs may also be noted. The
distinction between behaviors among the subgroups of cognitive
diseases is not well documented and should become a subject for
investigation rather than a matter of idiosyncratic clinical definition. There is an obvious need to study the relationship between behavioral cognitive and affective dimensions and biologic
changes throughout the natural course of the cognitive diseases.
Furthermore, the distinction between cognitive dysfunction as a
consequence of "normal" aging versus a pathological or accelerated process of impairment has not been adequately evaluated, in
our opinion.

There is enough of a data base, however, to define the following variants of cognitive diseases which for the moment remain
useful classifications for differential diagnosis. Manifestations
of cognitive dysfunction may reflect reversible disorders or permanent, i.e., nonreversible (structural or chemical), alterations
in the central nervous system (CNS). The nonreversible diseases
include senile dementia of the Alzheimer's type (primary neuronal
degeneration), senile dementia of the cerebrovascular type (multi-
infarct dementia), subcortical variants, other forms (Pick's

disease, Wernicke's-Korsakoff syndrome, Creutzfeldt-Jakob disease,
Huntington's chorea, multiple sclerosis). The differential diagnosis of cognitive diseases also requires an exclusion of the
reversible variants which include depression, the so-called acute
brain syndromes, or the pseudodementias, as well as normal pressure hydrocephalus.

Although cognitive impairment may be categorized in a number
of ways, we have distinguished two categories of cognitive diseases
for clinical purposes: The nonreversible variants (primary
neuronal degeneration, cerebrovascular, subcortical variants, and
other forms) and reversible conditions (acute brain syndromes and
depressions). A number of issues concerning this categorization
are worth noting. First, the so-called acute brain syndromes,
i.e., cognitive disturbance secondary to physical disease or
trauma, may or may not be of acute onset temporally; and in some
instances without proper treatment such disorders may lead to permanent structural damage. Second, it should also be carefully
noted that treatability is distinguishable from reversibility in
all of the disorders discussed. In the case of nonreversible disorders, the presence of a coexisting problem which may further
exacerbate cognitive difficulties is quite possible, and such conditions may reflect a component of the disorder which is reversible.
A third factor is the variability in apparent functional capacity
seen in the face of identifiable loss of brain substance. Even in
the presence of nonreversible cognitive diseases, decline in cognitive functioning is subject to wide variation. Environmental
and social deprivation and lack of emotional support appear to
accelerate the apparent decrement in functional capacity. Finally,
a variety of treatment strategies may play a role in modifying the
apparent course of cognitive impairment (Eisdorfer and Stotsky,
1977).

Reports on the prevalence of cognitive disorders in persons
65 years and older vary (Kay, 1977): 1.0 - 7.2% for "severe
dementias," 2.6 - 15% for "mild dementia," and 10 - 18% for combined rates, e.g., mild, moderate and severe. Furthermore, the
risk for cognitive disease shows a sharp increase with advancing
age. For persons in the eighth decade living at home in Newcastle
upon Tyne, the prevalence of brain syndrome was reported to be at
least 20% and probably is much higher. A category such as "mild
dementia" is a vague clinical term which has not been defined
psychometrically, and other data have shown that British and American psychiatrists tend to differ in their diagnosis of brain disorders and depression in the aged (Cooper, Kendall, Gurland,
Sharpe, Copeland and Simon, 1972; Kendall, Cooper, Gourlay, Copeland, Sharpe, and Gurland, 1972). U.S. psychiatrists diagnose
brain syndromes more frequently while their counterparts in Great
Britain note depression more frequently.

Despite such differences in reported prevalence rates, due in part to the type of survey populations surveyed (hospital vs. community), diagnostic criteria, number of disease categories (prevalence decreases as the number of categories increases), degree of severity used as a case criterion, duration of illness in patient at time of study, the amount of time invested in each case, and the method used to calculate prevalence rate (Gruenberg, in press), it does seem conclusive that the prevalence of cognitive disorders becomes more frequent with advancing age past 60.

Although it would be most valuable to refer to a natural history of the cognitive diseases in any discussion of differential diagnosis, we lack such data apart from clinical impressions. The psychometric literature reports low performance scores for cognitively impaired patients on the Wechsler-Bellevue (Botwinick and Birren, 1951; Cleveland and Dysinger, 1944; Dorken and Greenbloom, 1953; Halstead, 1943; Lovett-Doust, Schneider, Talland, Walsh, and Barker, 1953), the WAIS (Bolton, Brittin, and Savage, 1966; Kendrick, Parkboosingh, and Post, 1965; Kendrick and Post, 1967); Sanderson and Inglis, 1961), the Raven Progressive Matrices Test and the Mill Hill Vocabulary Scale (Hopkins and Roth, 1953; Kendrick et al., 1965; Kendrick and Post, 1967; Newcombe and Steinberg, 1964). Only one set of investigators have reported on the administration of an intelligence test to this population more than once (Kendrick and Post, 1967), but the retest interval was too short to provide useful information regarding the course of the disease. An approach to measuring the nature and course of the cognitive change will be discussed in a later section.

COGNITIVE DISORDERS: NONREVERSIBLE VARIANTS
ASSOCIATED WITH CNS DESTRUCTION

Senile Dementia of the Alzheimer's Type (Primary Neuronal Degeneration)

Alzheimer (1911) described a disorder in middle aged persons characterized behaviorally by loss of memory and orientation and cortical atrophy, loss of neurons, neurofibrillary tangles, granulo-vacuolar changes and plaques. Recent examinations of the brains of older persons with cognitive loss indicate that the most commonly observed alterations are primary neuronal changes whose etiology is unknown. The condition is probably most accurately referred to as Cognitive Dysfunction secondary to Primary Neuronal Degeneration. Currently, aluminum metabolism (Crapper, Krishnan, and Dalton, 1973), immune and autoimmune processes (Nandy, 1977; Eisdorfer, Cohen, and Buckley, in press) and latent viruses (Gajdusek, 1977; Traub, Gajdusek, and Gibbs, 1977) are a few areas

being investigated to clarify their possible role as etiologic
factors. For many years, cerebrovascular problems were incorrectly regarded as the primary cause of brain syndromes in the
elderly (Fisher, 1968; Paulson and Perrine, 1968). Current neuro-
pathologic data (Terry and Wisniewski, 1973, 1977) indicate that
only 15-25% of the true brain syndromes, i.e., nonreversible cognitive disease, are clearly due to cerebrovascular compromise.
Greater than 50% of the irreversible cognitive disorders are the
result of primary neuronal changes; the rest are mixed (with 8%
of those who manifest cognitive dysfunction showing no significant
morphological changes at autopsy).

     Although it has been customary to use the diagnosis of Alz-
heimer's disease if the age of onset occurred before age 65 and
senile dementia if similar cognitive dysfunction occurred after
the 65th birthday (Roth and Myers, 1976), careful clinical and
pathological analysis have not distinguished any differences
(Newton, 1948; Newmann and Cohn, 1953; Tomlinson, Blessed, and
Roth, 1970; Blessed, Tomlinson, and Roth, 1968; Terry and Wisniewski, 1977). Despite these findings the possibility exists that
presenile or early-onset Alzheimer's disease may be different from
late-onset senile dementia. Larsson and his colleagues (Larsson,
Sjögren, and Jacobsen, 1963) report that there is little commonal-
ity in the genetic (epidemiologic) predisposition to "senile dementia" and "presenile dementia." Not a single case of Alzheimer's
disease or Pick's disease was found among 2133 relatives (50
years and older) of 377 rigorously selected patient probands with
a diagnosis of senile dementia (arteriosclerotic dementia having
been excluded as much as possible), whereas the morbidity risk
for senile dementia was 4.3 times higher in the patients, children,
and siblings of the index cases compared to the general population.
However, since Larsson studied probands where the age of onset
varied between age 52 and 92, the stated younger age limit at least
argues for the inclusion of "presenile dementias."

     Larsson's data support the hypothesis that senile dementias
are inherited by an autosomal dominant gene with a prevalence of
40% at age 90. However, all the cognitive diseases are age-related
and it may be that the absence of a relationship with early-onset
Alzheimer's disease in Larsson's study is a function of the low
frequency of probands under age 65.

     Sjögren and his research team (1952) initially identified a
small series of 36 Alzheimer's patients (18 of whom were verified
histopathologically) and their family records. The morbidity
risk was 10% for parents and 3.8% for siblings of the index cases.
No senile dementia was observed in the family members. (The
total number of siblings and parents sixty years and older was
255).

Constantinidis and his laboratory team (1962) studied 67
patients. The risk of Alzheimer's disease alone was 1.4% in the
parents and 3.3% in the sibs. The combined risk for senile dementia and Alzheimer's disease was 10.6% for the parents and 9.5%
for the siblings. In both dementias, the risk figures for primary
relatives are higher than observed in the general population.
Furthermore, in spite of patho-anatomical similarities observed
between senile dementia, Alzheimer's disease, and "normal" older
persons, Larsson and his colleagues' observation that the absence
of Alzheimer's disease in the families of index cases with senile
dementia, and Sjögren's report of the absence of senile dementia
in the families of index cases with Alzheimer's disease, suggests
a different genetic basis for the two entities.

The proposition that there is a common basis for the presenile
and senile dementias has been advanced by Terry and Wisniewski
(1977). Some investigators consider Alzheimer's disease to be a
separate entity (Neumann and Cohn, 1953), others consider that
there are senile groups of presenile dementia (Delay and Brion,
1962), or consider senile dementia and Alzheimer's disease as one
and the same (Grunthal and Wenger, 1939; Arab, 1960; Tissot, 1968;
Albert, 1964). Pratt (1970) advocates a polygenic approach such
that the person is predisposed to develop either senile dementia
or Alzheimer's disease. Clearly, additional studies are indicated.
The age distinction can either be dropped or the disease can be
considered to have two variants, an early-onset and a late-onset
form.

The diagnosis of senile dementia with primary neuronal degeneration (the Alzheimer's variant) can be made when there is
evidence of progressive mental deterioration characterized initially by attentional problems and memory loss and eventually in
loss of orientation as to time, place, and person. In addition,
the possibility of a specific physical basis, a primary depression,
or a cerebrovascular compromise underlying the cognitive impairment
should be eliminated. However, the difficulties of early diagnosis in early-onset primary neuronal degeneration has been described by Nott and Fleminger (1975). Of 35 patients under age
65 diagnosed with senile dementia at York Clinic, only 15 (43%)
showed progressive cognitive impairment following the initial
diagnosis, whereas eighteen cases improved and two remained the
same (57%).

An examination of the Alzheimer's patient in the early stages
of the disease may reveal minor changes in affect and cognition.
Family members often have great difficulty pinpointing the time of
onset. The person afflicted often attributes changes in mental
functioning to other problems or illnesses. Lack of motivation
or loss of interest in once-important activities can be explained

as depression or feeling too tired to participate. Subtle changes
in personality such as marked anxiety or irritability may be the
first sign of Alzheimer's disease in individuals who recognize
that they have a difficult time organizing and keeping track of
their life activities. Furthermore, a number of other behaviors
are observed in patients with Alzheimer's disease state. These
include depression, anxiety, and paranoia, all of which may be
treatable.

In the later stages, patients with Alzheimer's disease can
show restlessness, perseveration, marked aphasia, affective lability, and eventually a total lack of knowledge regarding self and
relationship to family and friends as well as loss of bowel and
bladder control. However, the natural course of the disease has
not been well described. It is important to emphasize that although the patient gradually becomes more disoriented in the environment and tasks of daily living become increasingly difficult,
there is enormous variability in the extent and nature of the cognitive deficit, both within individuals and between individuals.

The electroencephalogram may show some generalized slowing of
the alpha rhythm and slow wave activity. However, focal abnormalities are uncommon in Alzheimer's disease in contrast to the
cerebrovascular brain syndromes. Obrist (in press) has reviewed
the scarce data on the use of the EEG to differentiate Alzheimer's
disease from the cerebrovascular variant associated with multiple
infarcts. The CAT (computerized axial tomography) scan may show
general cortical atrophy and/or enlarged ventricles in both Alzheimer's and the cerebrovascular variant. However, when the
latter is associated with multiple infarcts, a number of small
lucent areas can be seen in the CAT scan.

On autopsy, the brain will be smaller in size, show widened
sulci as well as the presence of neurofibrillary tangles, granu-
lovacuolar degeneration, and senile plaques. Blessed and his
colleagues (1968) have shown that beyond a certain minimum, the
density of plaque formation is positively correlated with the
extent of cognitive loss.

Cerebrovascular Variants

Persons presenting with symptoms of cognitive impairment (who
do not show evidence of a treatable physical cause or a primary
depression underlying the brain syndrome) can be classified as having brain syndrome of the vascular type if several of the following risk factors or signs are observed: (1) focal neurologic
signs; (2) signs and symptoms of retinal or brachial arteriosclerosis; (3) history of hypertension, diabetes, hyperlipidemia;

(4) history of black outs, fits or strokes, and (5) emotional
lability (Hall, 1976). Unless there is neurologic evidence or a
clear history of strokes (or a neuropathological report) it is difficult to diagnose a cerebrovascular variant with any degree of
certainty. Cerebrovascular brain syndrome may occur as a result
of a major stroke, multiple small strokes, vertebrobasilar insufficiency, carotid disease, or diffuse cerebrovascular disease
(Rivera and Meyer, 1975).

A detailed physical examination, including blood pressure,
pulse, ascultation, a search for bruits, as well as a detailed
neurological examination is necessary to identify objective signs
of multiple neurologic deficits which are probably due to vascular lesions. The presence of heart disease, abdominal aortic
aneurysm, and stenosis of a peripheral artery, i.e., evidence of
arteriosclerotic lesions, increase the probability of a vascular
etiology for a cognitive disorder.

At present, there are no data reported which distinguish differences between those patients with primary neuronal degeneration
or cerebrovascular disease. As a result, clinicians agree that it
is almost impossible to distinguish between them at a single point
in time in the basis of behavioral observations. Careful pathological analyses of the brains from persons with clear diagnoses
of arteriosclerotic dementia have shown profound softening of the
brain, usually involving more than 100 ml of tissue, whereas those
with "probable" diagnoses have shown softening in several areas
of the brain, involving between 50-100 ml of tissue (Tomlinson,
1970).

It has been suggested that patients with cerebrovascular
dementia show a more variable course with temporary partial remis-
sions followed by exacerbation. Where it exists, this fluctuat-
ing course makes the disorder particularly difficult to manage.
Families may have their hope for a cure reinforced only to see
them compromised in a series of crises.

Subcortical Variants

For many years mental impairment has been reported to accompany bilateral thalamic degeneration (Stern, 1939), olivoponto-
cerebellar degeneration (van Bogaert and Bertrand, 1929; van
Bogaert, 1946), and palladial degeneration (Winkelman, 1932). The
symptoms described are lack of interest, slowing of performance,
confusion, and agitation. Victor, Adams, and Collins (1971) have
also reported that the memory impairment observed in Wernicke's-
Korsakoff syndrome correlated with lesions in the thalamus, the
medial pulvinar, and the mamillary bodies.

The analysis of a variety of neurological syndromes, including progressive supranuclear palsy (Albert, Feldman, and Willis, 1974), characterized by subcortical pathology, suggest the following behavioral signs: (1) a memory deficit; (2) a slowing in cognitive performance; (3) affective changes, usually apathy and sometimes labile irritability; and (4) impaired learning. Albert and his colleagues have argued that these deficits are in contrast to the major language and perceptual-motor deficits observed in the cortical cognitive disorders.

Furthermore, there are some similarities between disorders of the frontal lobe (cf. Benson and Geschwind, 1975) and subcortical disease, e.g., affective lability, personality change, and abstraction deficits. Indeed, Albert and his colleagues have proposed a tentative hypothesis that the "common mechanisms underlying the subcortical dementias are those of impaired timing and activation" (Albert et al., 1974, p. 129). Pathological alterations in the thalamic and subthalamic nuclei effect the reticular activating system, impairing activities and slowing down performance. If subcortical cognitive diseases are induced by alterations in activation and timing mechanisms in the brain, then drugs which affect these parts of the brain may be useful intervention strategies.

Other Cognitive Disorders

Pick's Disease. Pick's disease is a rare cognitive disorder, inherited as an autosomal dominant disorder with high penetrance (Sjögren, Sjögren, and Lindgren, 1952). The diagnosis depends upon pathologic findings: symmetrical atrophy of the frontal, temporal, and posterior parietal lobes, the presence of Pick bodies and Hirano lesions, and the absence of neurofibrillary tangles and plaques (van Mansvelt, 1954; Neumann, 1949; Schocket, Lampert, and Lindenberg, 1968; Wisniewski, Coblenz, and Terry, 1972).

However, there are some clinical observations which may differentiate Pick's from Alzheimer's disease prior to an autopsy (Katzman and Karasu, 1975). Personality changes are reported to be initially more severe than memory impairment in Pick's disease, and attention seems to be markedly impaired relative to memory. Depression and anxiety are more rarely observed.

Wernicke's-Korsakoff Psychosis. These disorders, associated with alcoholism, are in reality nutritional diseases. Although the patient with Korsakoff's syndrome has an impaired memory and attentional abilities, he/she often functions at a stable, chronic

level of dysfunction for many years. This is in contrast to the
progressive deterioration in Alzheimer's disease. Indeed, many
aspects of personality remain quite intact. Some investigators
propose that there is a type of cognitive dysfunction associated
with chronic alcoholism that is different from Wernicke's or
Korsakoff's syndrome (Marsden and Harrison, 1972), although this
remains to be determined.

A combined Wernicke's-Korsakoff syndrome is described as a
total state of confusion, ataxia, oculomotor paralysis and, in
many cases, peripheral neuropathy. If the oculomotor symptoms and
ataxia improve, but not the confusion, after an injection of
thiamine intramuscularly, Wernicke's syndrome was probably present, regardless of the history of alcohol intoxification.

Creutzfeldt-Jakob Disease. Creutzfeldt-Jakob disease is a
rather rare cognitive disorder with a familial pattern of inheri-
tance; it is characterized clinically by generalized cognitive
dysfunction, myoclonus, and focal electroencephalographic activity
showing high voltage slow waves. The age of onset appears to be
earlier than Pick's or Alzheimer's disease, and the course is
rapidly progressive, associated with profound movement disorders,
usually lasting 8-12 months. For a detailed description of the
disease refer to Kirschbaum (1968, 1971); Traub et al. (1977);
Beck, Daniel, Matthews, Stevens, Alpers, Asher, Gajdusek, and
Gibbs (1969); and Gajdusek (1977).

Its etiology, an infection by a slow virus-like particle,
was first implicated when a filterable virus was isolated and
transferred to chimpanzees who in turn showed the characteristic
pathology (Beck et al., 1969; Gibbs, Gajdusek, Asher, Alpers,
Beck, Daniel, and Matthews, 1968). Now the gibbin as well as the
chimpanzee, a variety of new and old world monkeys, and the cat
have shown pathological signs indistinguishable from Kuru, a form
of the disorder reported among cannibals in New Guinea or the
more prevalent Creutzfeldt-Jakob disease upon transmission of the
virus (Lampert, Gajdusek, and Gibbs, 1972; Gajdusek and Gibbs,
1975; Beck et al., 1969). Manuelidis and colleagues (1975;
Manuelidis, Kim, Angelo, and Manuelidis, 1976) have transmitted
Creutzfeldt-Jakob disease from the human brain to guinea pigs,
and Gajdusek and his colleagues (1977) report a confirmation.
Furthermore, there have been reports of the possible transmission
of Creutzfeldt-Jakob disease from man to man (Duffy, Wolf, Collins, DeVoe, Steeten, and Cowen, 1974; Traub et al., 1974, 1975).

Gajdusek (1977) has reviewed the attempts to clarify the variety of diseases caused by a Creutzfeldt-Jakob disease (CJD) virus.
The pathological changes associated with slow virus infection

(neuronal vacuolation in the grey matter, reactive gliosis) have
also been observed in a variety of clinical entities, including
Alzheimer's and cerebrovascular variants (Traub et al., 1977).
Gajdusek posits that transmissible virus dementia (TVD) be used
to refer to the transmissible variants of cognitive diseases.

Huntington's Chorea. It is important to at least consider
the possibility of Huntington's in a differential diagnosis of
early-onset Alzheimer's disease. Abnormal involuntary movements,
lack of coordination, cognitive impairment, and affective lability
are frequently observed. Huntington's chorea usually occurs after
the third or fourth decade of life, sometimes lasting fifteen years.
The cognitive deterioration may appear before the characteristic
choreoform movements (Dewhurst, Oliver, Trick, and McKnight, 1969).
Huntington's chorea is under genetic control, inherited as a
single autosomal dominant disease with complete penetrance. Refer
to Barbeau, Chase, and Paulson (1973) and Caine, Hunt, Weingartner, and Ebert (in press) for a discussion of clinical and research issues.

Although Alzheimer (1911) did the first definitive study
based upon Huntington's (1872) description of the disease, Meynert
(1871) may have been the first to recognize that the most important
lesion site in Huntington's was the neostriatum where a loss of
neurons was observed. The bilateral atrophy of the neostriatum
and the cerebral cortex is really the effect of a progressive metabolic genetic defect that is acting long before the neurons de-
teriorate and die. The neuropathology has been well described
(Earle, 1973). The most characteristic pathologic signs are observed when the brain is cut in a coronal section: bilateral and
symmetrical atrophy of the caudate nucleus, and putamen, as well
as the globus pallidus.

Current research is emphasizing clinical assessments of abnormal muscle movements, neuropsychological testing, biochemical
assays, and genetic studies (linkage and chromosomal analyses) to
detect preclinical Huntington's disease. With the older adult presenting cognitive dysfunction, a query regarding family genetic
history, early adult age of onset, and the presence of abnormal
muscle movements provides useful diagnostic information. Genetic
counseling appears to be the approach of choice in the control
of this disorder.

Miscellaneous Cognitive Diseases. Cognitive impairment can
occur in Parkinson's disease, especially when it is associated
with arteriosclerosis as well as the genetically determined neuronal storage diseases (Kuf's disease, Tay-Sach's disease),

the adult leukoencephalopathies (multiple sclerosis, Binswanger's encephalopathy) and the miscellaneous neurofibrillary diseases such as the amyotrophic lateral sclerosis-Parkinsonian dementia complex of Guam (cf., Slaby and Wyatt, 1974).

REVERSIBLE COGNITIVE DYSFUNCTION

Acute Brain Syndromes

In a person with uncomplicated chronic brain syndrome or cognitive dysfunction, it is uncommon to see overwhelming anxiety, florid delusions, or hallucinations. When these occur or when the onset of cognitive dysfunction is rapid, it is important to probe for a physical cause underlying the behavioral change, i.e., acute organic brain syndrome. A patient with a chronic brain syndrome who then develops a superimposed acute brain syndrome presents a diagnostic problem for the clinician. Superficially, this change might herald a progression of the chronic disease. However, if there is an underlying physical cause it may be remedial. Even a chronic brain syndrome can have a correctable underlying cause, and it is inappropriate to dismiss the disorder as untreatable. Refer to Libow (1977) for a detailed discussion of clinical medical investigations which are valuable in the diagnosis of the cognitively impaired elderly.

The causes of "acute" brain syndrome may be classified into at least six general classes as described below:

Structural Causes. About half of the patients with intracranial tumors show mental changes. The possibility of a subdural hematoma resulting from a trauma which may not be remembered by the patient should be investigated.

Normal pressure hydrocephalus, a recently described syndrome, may be corrected, if present, by relieving the increased cerebral spinal fluid (CSF) in the ventricles with a surgical shunting procedure. In normal pressure hydrocephalus, a communicating hydrocephalus (identified by pneumoencephalography, CAT scan, or an abnormality in CSF absorption) is reported; this disorder is associated with a disturbance in gait, cognitive impairment, and urinary incontinence, usually late in the stages of the disease (Benson, 1974; Hakim and Adams, 1965; Katzman and Pappius, 1973; Katzman and Karasu, 1975). The differential diagnosis between Alzheimer's disease and normal pressure hydrocephalus is very difficult. One major criterion for the diagnosis is response to shunt therapy (Katzman and Karasu, 1975). However, this is a

rather profound approach. Of some concern for the future may be
the need for a more precise basis for the diagnosis. Without such
precision the possibility exists that we will shortly witness a
major increase in the number of unnecessary neurosurgical procedures based upon questionable diagnoses of this disorder.

   Drugs. The pharmacokinetics of many drugs would indicate
that older persons may be particularly sensitive to drug effects
as well as to their side effects. As a result, a wide variety of
prescribed drugs ranging from diuretics to psychotropic medication
may have untoward side effects including cognitive impairment.
Misuse of prescription drugs among the aged is common (Hemminki and
Heikkila, 1975) and complicates an already difficult situation involving use of multiple physicians often in poor communication,
the use of drugs "inherited" from a neighbor, the use of outdated
drugs, or the substitution of one medication for another. Indeed,
drug related problems are a major source of difficulty in the
aged.

   The use of sedatives, both prescribed and ones that can be
obtained over-the-counter, is a common cause of impaired cognition. Sedative hypnotic agents through their general action of
general cortical depression can mimic or exacerbate symptoms of
cognitive disorder. Social and behavioral functioning may improve
dramatically when these drugs are stopped.

   The use of antipsychotic, tricyclic antidepressants, and
anti-Parkinsonian agents, all of which have potent central anticholinergic side effects present a more subtle problem. The
anticholinergic side effects may be very strong and the patient
may show what appears to be a typical brain syndrome. The treatment again is the halting of medication. The hypotensive effect
of antipsychotic medications may lead to secondary problems.
Diuretics may upset sodium and potassium balance, and this will
lead to cognitive problems.

   Older people may be very sensitive to bromides and in levels
which at one time were considered therapeutic for anxiety. They
may become either delirious or present an array of psychiatric
problems ranging from paranoid psychosis to the appearance of a
cognitive disorder. Bromides are still in several over-the-counter
preparations and may also be in prescription medications.

   The list of possible drug problems is extensive and suggests
that drug history be carefully reviewed. Indeed, drug-free periods, during which all medications not felt to be needed to support life are gradually withdrawn, are important before a diagnosis
is made. (See Chapter 9 for a more extensive discussion of the
use of chemotherapy with older adults.)

Hypoglycemia. Hypoglycemia is a common cause of acute brain
syndrome and it can exaggerate the severity of a chronic brain
syndrome in persons who have been on what appears to be a reasonable diabetic program with either a long-acting oral hypoglycemic
or insulin with an inadequate caloric intake or reduced hepatic
or renal function. Vascular accidents and "progressive dementias"
may be eliminated by adjusting the diabetic program, insuring that
food intake is adequate.

Congestive heart failure. The occurrence of congestive heart
failure can also impair cognition in an individual with marginal
or more advanced cognitive dysfunction. Congestive heart failure
should be pursued zealously since it is often easy to medically
reverse the situation.

Infections. Acute and long-standing infections may cause
intellectual deterioration which will reverse when the infection
is treated. Parenthetically, a disease often overlooked in the
elderly is syphilis. It may exist in the older population as
tertiary syphilis, but there is documentation of recent infection
in sexually active older people (Raskind and Eisdorfer, 1976b).

Metabolic and Nutritional Causes of Acute Brain Syndrome.
Metabolic causes often underly acute brain syndrome. Such endocrine problems as thyroid insufficiency are among the most common
diseases. Throidectomy was once common and such people have often
been lost to follow-up by their initial physician. If thyroid
replacement is neglected, the individual can drift into a state of
thyroid insufficiency or myxedemia. This presents as lassitude,
irritability, and cognitive dysfunction. Thyroid insufficiency
is treatable and should not be overlooked.

Persons with pernicious anemia can show symptoms of organic
brain syndrome and psychiatric problems in the absence of gross
blood abnormalities. If vitamin B-12 deficiency is present, this
is another treatable cause of the apparent brain syndrome deterioration, though long delays in treatment will result in permanent
CNS alterations.

Elevated calcium produces apathy and confusion; hypercalcemia
occurs in a number of disease states in late life, e.g., lung and
breast cancer, multiple myeloma, and hyperparathyroidism. Hypo-
calcemia observed in renal failure or hypoparathyroidism (usually
related to thyroid surgery) may also present as a brain syndrome
and may be overlooked. Persons with hypernatremia and hyponatremia
(high and low serum sodium respectively) also may demonstrate

confused and disoriented behavior. Indeed, many traumata and infections may be implicated in this problem.

DEPRESSIVE DISORDERS

Symptoms of depression often accompany a variety of diseases
in the elderly, including endocrine disturbances, metabolic disorders, neurological diseases, and the chronic nonreversible brain
syndromes (Salzman and Shader, 1977). Theoretically, the chronic
brain syndromes and depression are quite different, but in clinical
practice they may be difficult to separate. The psychomotor slowing as well as other behaviors observed in depression are often
incorrectly diagnosed as nonreversible brain syndrome and physical
frailty.

Raskind (1976) has reported that 9% of those persons referred
to the Older Adults Outreach Program (Raskind, Alvarez, Pietrzyk,
Westerlund, and Hurland, 1976) with a cognitive disorder, e.g.,
diagnosis of senile dementia, were in fact manifesting a clinical
depression. An additional 21% referred with the same diagnosis
had a reversible cognitive problem associated with the wide range
of medical disorders (acute brain syndrome) described above.

In most instances, depression and acute brain syndromes are
treatable or even preventable illnesses. While a depression may
be difficult to detect immediately, persistant and careful interviewing will clarify its presence (Post, 1975). If the latter is
not treated promptly and effectively, permanent brain damage may
result. Furthermore, an acute brain syndrome can be superimposed
on a chronic brain syndrome and should not be regarded as untreatable.

It is not unusual for a primary depression to occur abruptly.
In the case of a patient presenting a rapid onset of cognitive
and affective symptoms, the elimination of a physical cause is
the mandatory first step. In the more usual case of the gradual
onset of symptoms, vigorous assessment of depressive affect and
cognitive impairment is imperative. In the future the development
and application of cognitive diagnostic tools may prove useful to
the differentiation of depression and cognitive impairment
(Crookes and McDonald, 1972; Kendrick, 1972; Miller and Lewis,
1977). Miller and Lewis (1977) have developed a spatial memory
recognition task in which older persons with depression did not
show an impairment in memory processing in contrast to those with
brain syndrome. However, the former appeared to use more conservative strategies in the task compared to the brain syndrome sample.

Depression per se does not appear to increase the risk for chronic cognitive diseases. Nonreversible brain syndrome occurs with the same frequency in depressed elderly population as in the general older population (Kay, 1962; Kay, Roth, and Hopkins, 1955; Post, 1972; Roth, 1955). However, depression is frequently observed in the patient with cognitive disease and may be more prevalent in the cerebrovascular variant relative to the Alzheimer's variant (Post, 1962, 1975). In any event, depression should be actively treated. Refer to Eisdorfer and Stotsky (1977) for a discussion of treatment and rehabilitation strategies.

A variety of depressive disorders are observed in the elderly. These include reactive depressions, unipolar depression, bipolar (manic-depressive) disorders and existential sadness (Eisdorfer, 1977; Verwoerdt, 1976). A family history and a personal history of the periodicity and direction of mood swings is of major value in making the diagnosis. This is particularly the case where stressful life events may be implicated in the precipitation of a reactive depressive disorder.

Depression is perhaps the most common functional psychiatric problem in the elderly but is associated with measurement difficulties, and for this reason epidemiological data are difficult to interpret. Gurland (1976) has discussed the age-related prevalence rate of depression in terms of both clinical diagnoses of depression and age patterns of depressive symptoms. Depressive disorders as diagnosed by psychiatrists are most frequently observed between the ages of 25 and 65, and probably decline after age 65. The mild neurotic depressives are more frequent after age 40. Although women have higher rates of depression than men through midlife, the sex differences become smaller between 45 and 65, and may even favor males after age 65. However, in studies where symptoms are analyzed, age groups above 65 have the highest rates of depression. The frequency and manifestation of depression in the aged remain important areas of investigation for research. Treatable cases of depression in the elderly may be overlooked or misdiagnosed as anxiety neurosis, a brain syndrome, or a physical disease.

The symptoms of depression are wide ranging. Somatic complaints and hypochondriasis can mask depression in the older patient. A depressed mood or chronic state of sadness is associated with changes in the face, gait, posture, grooming, and general slowing of responsivity. Physical symptoms, e.g., loss of weight, loss of appetite, insomnia (particularly early awakening or so called terminal insomnia), as well as physical complaints such as chronic pain without a clear cut etiology, constipation, feelings of weakness or lack of energy, can be observed in depression. Common psychological complaints are feelings of hope-

lessness, helplessness, worthlessness, and chronic feelings of
inadequacy. Often emptiness and "nothingness" and the absence of
any feelings may be reported. Blumenthal (1975) has pointed out,
however, that older people may be frail and have physical problems
and movement difficulties apart from depression. Thus the validity of scales such as the Zung (1965) and Hamilton (1960) which
emphasize the importance of physical symptoms as a manifestation
of depressive state is doubtful.

Depression may also be characterized by agitation and behavior
such as hand wringing and self-mutilation attempts, or feelings of
guilt with a goal of seeking punishment. Sometimes guilt appears
as reminiscences of early events which seem to plague the individual
as obsessional thoughts. Agitated depressions are a management
problem. They are a hazard because such individuals can in a fit
of agitation act out, fall and fracture their hips or sustain head
injuries. Vigorous intervention may therefore be needed.

Physical illness and alcoholism are important etiologic factors in depressive behavior. Suicide, a significant risk in elderly depressed men, is often associated with depression and
alcohol intake (Resnick and Kantor, 1970). An individual with a
history of depression, high alcohol consumption (or use of drugs
such as central nervous system depressants), recent "bad news,"
and a previous attempt of suicide shows a heightened potential for
self destructive behavior.

A number of precipitants of depressed behavior should be considered: the loss of a loved one, loss of status, prestige, material possession, physical trauma, loss of physical competency and
mobility, humiliation, social embarrassment, or feelings of inadequacy. Often the mere threat of such loss is sufficient to act
as a cue. Older persons, like the young, have reactive depressions and for these problems social and psychological support are
the treatment of choice. The multiple losses experienced by
older persons may in fact lead to a cumulative reactive depression.
However, there is little hard data regarding the effect of multiple stressors on psychiatric symptoms in the elderly, although
life event changes have been reported as a precursor of disease
in the general population (Holmes and Rahe, 1967).

Depression can also be confused with unhappiness or existential sadness which may be appropriate to the situation in which
the elderly patient finds herself or himself. Sadness in an individual may be indicative of an existential state as well as of a
disease. There are many reasons to be sad in the later decades,
e.g., reminiscences, social and financial deprivation associated
with aging, loss of opportunities. Existential sadness should be
distinguished from depression however. The former might be more

appropriately managed by allowing the individual involved to experience and work through his/her feelings rather than to attempt
intervention with drugs and special supports. Helping the individual change a social situation or develop an improved socio-
political economic state might be far more appropriate than any
patient oriented treatment since the older person is not a patient.
In situations where need is evident, direct social or financial aid
may be needed. What is not needed is to convert an individual re-
acting appropriately to an unhappy situation to a "patient" or
client where that role transforms the individual into a passive
recipient of professional care or services. Clinical depression
may exist in such instances, however, and should be considered if
symptoms last more than six weeks, if weight loss is significant,
if sleep is disrupted, if the degree of sadness seems to exceed
the losses sustained, if worthlessness and guilt are expressed and
the sadness is getting worse.

If physical illness, alcoholism, and losses can be eliminated
as possible precipitants of depression, then one should consider
the possibility of unipolar depression characterized by a history
of previous depressive disorder (often successfully treated) or
bipolar (manic-depressive) illness where depressive episodes may
be interspersed with manic episodes. These latter are characterized by hyperactive, elated or agitated states and a family history
of similar illness and/or depression (Winokur and Clayton, 1967;
Winokur, Clayton, and Reich, 1969).

Once a physical cause of the depression has been ruled out,
the depression can be identified as an illness which should be
treated (Raskind and Eisdorfer, 1976a; Friedel and Raskind, 1977).
A useful clinical assumption is to recognize that in depression
one is not dealing with just one therapy but with a therapeutic
armamentarium (Eisdorfer and Friedel, 1977). Psychosocial supports are the method of choice in a reactive depression. Lithium
is valuable in the case of a bipolar depression. Tricyclics appear to work in most unipolar depressions. Schizo-affective
disorders are related to depression and often are treatable. In
summary, it appears that precision in observation and diagnosis
pays off in greater precision in treatment and prognosis.

PARANOID BEHAVIOR

Typically, paranoia is a symptom, not a disease, in which the
person has delusions of persecution or grandeur. Many different
diseases can produce paranoia. Since paranoid behavior may occur
in the absence of any cognitive loss, this may be a very important
differential. Rapidly progressive brain syndrome may first present in the form of paranoia or depression.

   Kay and his colleagues (Kay, 1963; Kay and Roth, 1961; Roth
and Morrissey, 1952) have described a syndrome of paranoid behavior
in the aged characterized by an extensive delusionary system often
associated with hallucinations but without any signs of cognitive
loss, i.e., orientation as to time, place, and person as well as
attention are intact. This disorder is referred to as paraphrenia
by the British. Its prevalence in the community is not clearly
defined, although Raskind (1976) reported that in a group of patients referred to his crisis intervention program for older persons about 4.5% of the sample were primarily paraphrenic.

   The paranoid individual appears superficially cooperative,
yet is basically very resistant to treatment and often noncompli-
ant with antipsychotic drug therapy. Earning the patient's trust
and understanding that prescribed medication is helpful is as important and as basic to the chemical management itself. It is of
great interest that older paranoid patients, especially when they
have developed their paranoid and schizophrenic-like illness in
old age, are still very receptive to helping agencies. They welcome a doctor or a nurse as long as the shared goal is medical
care.

   It is important not to stop a diagnostic work-up prematurely.
An effective interview may be supportive and informal. If the
paranoid individual is allowed to talk without confrontation and
criticism, the paranoid process may suddenly emerge and become
quite florid within a few minutes. It is particularly important
to press for answers to a mental status evaluation in these patients who often are quite clever in evading answers to direct
questions. Possible problems with hearing and memory should also
be evaluated since such loss may present with a paranoid-like
flavor.

   In most cases paranoid symptoms can be treated. The symptomatology is responsive to medications as well as interpersonal
therapy and social supports. In the patient with paranoia, in
addition to a brain syndrome, cognitive impairment may appear to
improve when the paranoid behavior and attendant anxiety are
reduced.

   An important issue to be considered in the patient with paranoia but without any clear evidence of cognitive deterioration, is
a careful evaluation of treatment strategy. The presence of
paranoia does not automatically require that institutionalization
be sought for the patient. Treatment and disposition strategies
should consider how disturbing the problem is to the patient and
how disturbing the individual is to the surrounding environment.
Somatic treatment should be used cautiously in situations when
the paranoia is mild, although it may be most helpful for the more

disturbed patient who has compromised his/her status in the community. It is an error to rely too heavily upon chemical treatment without looking at how the environment or the people in the environment tolerate the patient. Medications often have negative side effects. Often the patient may be more resistant to the medication at first but at a later point in time, more receptive. Finally, where medication is indicated but compliance is a problem, the use of depot (long acting injections) may be more effective. This is particularly the case where patients are living at home and alone with no one to supervise their care. Much reduced doses are indicated in such instances.

ASSESSING THE INDIVIDUAL PRESENTING SYMPTOMS OF COGNITIVE DYSFUNCTION

There are certain fundamental steps in the differential diagnosis of the cognitively and emotionally impaired elderly: (1) a careful physical, psychiatric and neurologic examination; (2) clinical laboratory tests; (3) consideration of elective laboratory tests (radiologic and radioisotope studies); (4) health history and interview with the patient, family members, or significant others; (5) psychological testing, including cognitive assessment.

Medical Examination

Older persons may be hesitant to discuss physical and emotional complaints because the symptom is expected as a natural course of aging or because of embarrassment, e.g., rectal pain, sexual problems. Furthermore, the presenting problem for many underlying psychological disturbances may be in somatic dysfunction, e.g., appetite, pain, sleep disturbances, and, of course, physical dysfunction may lead to impaired behavior.

A thorough physical examination should be done to evaluate the health of major systems. A general examination including pulse, blood pressure, auscultation of the heart and lungs, and a search for arterial bruits is important in order to evaluate the possibility of a cerebrovascular etiology to the cognitive impairment. An neurologic examination of cranial nerves, motor, sensory and cerebellar systems, gait, frontal lobe reflexes as well as mental status (orientation, calculation, speech, abstraction, and memory) will help evaluate focal neurologic lesions.

Clinical Laboratory Tests

A series of basic laboratory tests are recommended to evaluate the metabolic status of the individual. These should include electrolyte studies, tests of endocrine and hepatic function, as well as other measures of physical status.

Radiologic Studies

Computerized axial transverse tomography (CAT scan) is useful to diagnose cortical atrophy (large sulci and ventricles) associated with nonreversible brain syndromes. The availability of the CAT Scan has reduced the need for pneumoencephalography in the aged. It is important to emphasize that findings of atrophy on the CAT scan are interpretable as Alzheimer's disease only with a clinical picture of cognitive impairment. Increased ventricular size has been reported in the brains of healthy people after age 70 (Barron, Jacobs, and Kinkel, 1976).

The CAT scan also is useful in the diagnosis of cerebrovascular variants of brain syndromes associated with multiple small cerebral infarcts; small lucent areas, indicating infarcts, are seen in addition to the large ventricles. Huckman, Fox, and Ramsey (1977) have reviewed the usefulness of computerized tomography in the diagnosis of degenerative diseases of the brain.

Radioisotope Studies

A brain scan as well as regional cerebral blood flow measurements are elective tests used to establish the degree of cerebral blood flow impairment. Cerebral blood flow as well as oxygen uptake in the brain are reduced in patients with nonreversible cognitive diseases (Simard, Olesen, Paulsen, Lassen, and Skinhoj, 1971; Olesen, 1974). Gustafsen and Risbery (1974) showed that reduced cerebral blood flow in the posterior temporal, parietal, and occipital regions were associated with poor memory performance and severity of aphasia. Isotope cisternography is important in a diagnosis of normal pressure hydrocephalus.

The History

A careful history from a relative or a close associate is critical in order to evaluate the time and type of onset (rapid vs. slow onset, slow progressive vs. stepwise deterioration), medical problems, and difficulties in activities of daily living. Information regarding work accomplishments and educational achievements are useful to assess premorbid functioning. A family history is important to the evaluation of a genetic component in the cognitive disorder or chorea, if present. Finally, a special effort should

be made to gather information about the patient to provide material
to test episodic memory, e.g., memory for that person's important
life events.

The Patient Interview

The interview with the cognitively impaired patient presents
special problems. It is important to speak slowly and distinctly
and to determine whether the patient may be having difficulty
processing questions because of hearing or visual losses. If the
patient is a resident in a long-term care facility, interviewing
in a staff conference can be helpful for both the patient and the
staff. The interview by a visiting psychiatrist or psychologist
provides the opportunity for the staff to get a different view of
a patient with a brain syndrome, who may present to them as a complainer, hypochondriac, or other behavior problem.

The older resident with cognitive impairment is usually known
to the staff by a variety of behavior problems. The older person
often has difficulty asserting some control over the environment
or in making specific needs known, and it is common for the patient
to regress to a behavior pattern that was successful in earlier
life. This may easily be interpreted as a deterioration in performance rather than a reversible behavior problem. Furthermore,
the social and interpersonal style of some patients may be elusive.
Humorous or smart remarks may characterize vague responses, but it
is important to make the effort to have the individual give the
best answers to questions.

The Cognitive Evaluation

A cognitive evaluation is a critical component of the evaluation process. Current mental status examinations and standard
psychological test batteries are not optimal instruments to evaluate the nature and course of cognitive dysfunction observed in the
impaired aged. Although they provide an important gross assessment
of cognitive performance level, they do not provide measurement of
discrete cognitive units or processes that may be useful to characterize disturbances in brain functioning. For example, word
learning tests are very useful to the diagnosis of chronic nonreversible cognitive diseases in the elderly (Miller, 1977), but it
is difficult to relate the findings to current theories of cognitive processes. It would be helpful to use a technical approach
that has diagnostic validity and also permits a theoretical analysis of cognitive deficit.

Previous clinical diagnostic work-ups have relied heavily

upon a variety of brief mental status examinations (cf. review
Salzman, Kochansky, and Shader, 1972). The most frequently used
are variations on a ten question examination to assess orientation
in time and place, recent and remote memory (Kahn, Goldfarb,
Pollack, and Peck, 1960; Pfeiffer, 1975). A 22-item dementia rating scale for cognitive and personality functioning (Blessed et
al., 1968) has proven useful in British studies (Kay, 1977). It
is also of interest that ratings of severity of dementia on this
latter scale correlated with mean plaque count at autopsy (r =
0.77).

Although a rapidly administered mental status (MS) questionnaire is a primary diagnostic tool, the usefulness of present forms
for measuring change is limited. In practice, busy physicians
and other health professionals usually limit their assessment of
mental status to orientation in time, place and person, perhaps
the three most insensitive indicators of brain syndromes (Jacobs,
Bernhard, Delgado, and Strain, 1977). Mental status examinations
seldom evaluate other variables that mediate performance levels,
especially anxiety and depression. Another basic flaw of standard
mental status examinations is that they fail to measure an essential aspect of cognitive dysfunction, its variability.

A number of investigators have attempted to develop better
clinical instruments for assessing the existence of diminished
cognitive capacity (Jacobs et al., 1977; Folstein, Folstein, and
McHugh, 1975). Restructuring mental status examinations to take
advantage of theoretical and empirical advancements in cognitive
psychology and neuropsychology should be useful to gather data in
order to clarify and differentiate cognitive diseases.

Several authors (Cohen and Eisdorfer, 1977; Crook, 1977;
Schaie and Schaie, 1977) have summarized a number of important
issues in the clinical evaluation of the impaired elderly. The
challenge to the clinician is to evaluate specific as well as
general cognitive deficits, variability in the particular behaviors, the effect of motivation and personality upon cognitive
performance, the extent of the limitation of auditory, visual and
motor losses on deficit performance, and the potential for training, rehabilitation, or management. Since most of the sixty or
more psychometric tests used with the elderly lack norms and
validity data (Crook, 1977), their naive use with the older (impaired) adult can be harmful rather than helpful.

As indicated earlier, examination of WAIS scores show the
expected poorer performance of older persons with the chronic
nonreversible brain syndromes, but reveal little about the nature
or maximal extent of the deficit. Since further psychometric
testing is limited by the lack of appropriate and valid test

instruments, an alternative approach is to evaluate efficiency of the person's information handling abilities in a variety of experimental cognitive paradigms. The few studies conducted with largely unselected groups of impaired elderly suggest deficits in attention, learning, and memory. The most systematic studies of learning and memory have been done by Inglis and Miller (cf. review, Inglis, 1970; Miller, 1977). A review of language disorders found in the brain syndromes has recently been prepared by Albert (1977). However, none of the studies to date have attempted to clarify the natural course of cognitive dysfunction, which may involve differential decline of cognitive traits. For example, Miller and Hauge (1975) showed that a sample of patients with early-onset brain syndrome did not demonstrate a loss in vocabulary, a finding which has been reported by other investigators (Critchley, 1964; Stengel, 1943, 1964), who tested mixed populations.

A major issue is the difference between the measurement of deficit in a generalized way and the assessment of adaptive functioning in more specific modalities. Eisdorfer and Cohen (in press) observed significant differences in verbal learning performance associated with differing autonomic nervous system (ANS) state in two subgroups of patients with Alzheimer's disease, who were otherwise similar in age, duration of illness, and test scores on the WAIS and Wechsler memory scale.

At the moment, with the possible exception of a few verbal learning tasks (Caird, Sanderson, and Inglis, 1962; Kendrick et al., 1965; Walton and Black, 1957; Walton, White, Black, and Young, 1959), no existing cognitive and neuropsychological tests have been established as valid and reliable diagnostic tools which show minimal to moderate differences among the impaired aged. However, there are a wide variety of paradigms that could potentially be refined, adapted, and doubtless standardized for such purposes. Cognitive studies in patients with nonreversible brain syndromes (Cohen and Cox, in press; Miller and Lewis, 1977; Talland, 1965), Korsakoff's syndrome (Cermak, Butters, and Goodglass, 1971) and Huntington's chorea (Caine, Ebert, and Weingartner, in press; Weingartner, Caine, and Ebert, in press) are examples of clinically relevant experimental analyses of cognitive dysfunction.

A detailed cognitive evaluation is obviously important and some general suggestions for evaluation can be made in spite of the current lack of appropriate test instruments. The WAIS is useful as a general test of intellectual functioning unless the subject is to be tested frequently. A neuropsychological assessment should include an evaluation of language, e.g., Boston Diagnostic Aphasia Examination, specific tests of parietal lobe functioning (drawing to command and to copy, right-left orientation), frontal lobe function (motor sequencing tasks, tapping) as well as subcortical functioning (paced auditory serial addition). The Halstead-Reitan (Reitan and Davison, 1974) is potentially a valuable resource but probably requires modification and standardization before it can be useful with the older (impaired) adult. Further discussion of the Halstead-Reitan battery as applied to elderly persons is found in Chapter 3.

The utility of word learning tests in diagnosis suggests that a systematic analysis of performance levels using a series of cognitive test paradigms could be a productive approach to developing a differential cognitive assessment technology (Cohen and Eisdorfer, 1977). A cognitive deficit could occur in attending to incoming material, acquisition, or the encoding and forgetting of material in short-term as well as long-term memory. It has been argued by some theorists that a human being can be visualized as an information handling system. Components of such a system have been identified and it is possible to evaluate the efficiency (or inefficiency) of those components. For example, one could study: (1) sensing, the quality of transmission by sensory organs; (2) pattern recognition, the initial coding of sensory information and the coding of information at the sensory-memory interface; (3) language, recognition of linguistic meanings; (4) remembering, the nature of memory loss as well as the organization of memory; or (5) reasoning, processes of making inferences (Rumelhart, 1977). A number of models of information handling have been proposed and all have their critics. However, key theoretical concepts and empirical findings remain to be used to improve our understanding of cognitive impairment in the elderly.

In summary, our present tools fall short of evaluating the cognitive processes affected in the senile dementia, the extent of the deficits, the availability of alternate capacities, the effect of and the ability to control environmental demands, as well as changes in behavior. The end result may be a program of retraining tailored to the individual, to mobilize his or her available skills and assets.

REFERENCES

Albert, E. Senile demenz und Alzheimer'sche krankheit als ausdruck des gleichen krankheitsgeschehens. *Fortschritte der Neurologie, Psychiatrie und Ihrer Grenzgebiete*, 1964, 32, 625-673.

Albert, M. Language in the aging brain. Paper presented at Conference on Cognition and Aging, Seattle, January, 1977.

Albert, M. L., Feldman, R. G., and Willis, A. L.   The subcortical
    dementia of progressive supranuclear palsy. _Journal of_
    Neurology, Neurosurgery & Psychiatry, 1974, 37, 121-130.
Alzheimer, A.   Uber eigenartige krankheitsfälle des späteren
    Alters. _Zeitschrift fur der Gesamte Neurologie und Psychi-_
    atrie, 1911, 4, 356-385.
Arab, A.   Unité nosologique entre démence sénile et maladie
    d'Alzheimer d'après une étude statistique et anatomo-
    clinique. _Sistema Nervosa_, 1960, 12, 189.
Barbeau, A., Chase, T. N., and Paulson, G. W. (Eds.). _Huntington_
    Chorea, 1872-1972. _Advances in Neurology_, Vol. 1. New
    York: Raven Press, 1973.
Barron, S. A., Jacobs, L., and Kinkel, W. R.   Changes in size of
    normal lateral ventricles during aging determined by com-
    puterized tomography. _Neurology_, 1976, 26, 1011-1013.
Beck, E., Daniel, P. M., Matthews, W. B., Stevens, D. L., Alpers,
    M. P., Asher, D. M., Gajdusek, D. C., and Gibbs, C. J., Jr.
    Creutzfeldt-Jakob disease: The neuropathology of a transmis-
    sion experiment. _Brain_, 1969, 92, 699-716.
Benson, D. F.   Normal pressure hydrocephalus: A controversial
    entity. _Geriatrics_, 1974, 29, 125-132.
Benson, D. F., and Geschwind, M.   Psychiatric conditions associ-
    ated with focal lesions of the central nervous system.   In
    M. F. Reiser (Ed.), _American Handbook of Psychiatry_. New
    York: Basic Books, 1975.
Blessed, G., Tomlinson, B. E., and Roth, M.   The association be-
    tween quantitative measures of dementia and of senile change
    in the cerebral grey matter of elderly subjects. _British_
    Journal of Psychiatry, 1968, 114, 797-811.
Blumenthal, M. D.   Measuring depressive symptomatology in general
    population. _Archives of General Psychiatry_, 1975, 32, 971-
    984.
Bolton, N., Brittin, P. G., and Savage, R. D.   Some normative data
    on the WAIS and its indices in an aged population. _Journal_
    of Clinical Psychology, 1966, 22, 184-188.
Botwinick, J., and Birren, J. E.   Differential decline in the
    Wechsler Bellevue subtests in the senile psychoses. _Journal_
    of Gerontology, 1951, 6, 365-368.
Bremer, F.   A social psychiatric investigation of a small commun-
    ity in northern Norway. _Acta Psychiatrica et Neurologica_
    Scandinavica, 1951, 62 (Supplement 1).
Busse, E. W., Dovenmuehle, R. H., and Brown, R. G.   Psychoneurotic
    reactions of the aged. _Geriatrics_, 1960, 15, 97-105.
Busse, E. W., and Pfeiffer, E.   Functional psychiatric disorders
    in old age. In E. W. Busse and E. Pfeiffer (Eds.), _Behavior_
    and adaptation in late life. Boston: Little Brown, 1975.
Caine, E. D., Ebert, M. H., and Weingartner, H.   An outline for
    the analysis of dementia: The memory disorder of Hunting-
    ton's disease. _Neurology_, in press.

Caine, E. D., Hunt, R. D., Weingartner, H., and Ebert, M. H. Hunt-
    ington's dementia: Clinical and neuropsychological features.
    Archives of General Psychiatry, in press.
Caird, W. K., Sanderson, R. E., and Inglis, J. Cross validation of
    a learning test for use with elderly psychiatric patients.
    Journal of Mental Science, 1962, 108, 368-370.
Capel, W. C., and Stewart, G. T. The management of drug abuse in
    aging populations: New Orleans findings. Journal of Drug
    Issues, 1971, 1, 114-120.
Cermak, L. S., Butters, N., and Goodglass, H. The extent of mem-
    ory loss in Korsakoff patients. Neuropsychologia, 1971, 9,
    307-315.
Cleveland, S., and Dysinger, P. Mental deterioration in senile
    psychosis. Journal of Abnormal and Social Psychology, 1944,
    39, 368-372.
Cohen, D., and Cox, G. Memory for a geometrical configuration in
    senile dementia. Experimental Aging Research, in press.
Cohen, D., and Eisdorfer, C. Cognitive theory and the assessment
    of the impaired elderly. Paper presented at Geriatric
    Assessment Workshop, Los Angeles, April, 1977.
Constantinidis, J., Garrone, G., and D'Ajuriaguerra, J. L'hérédité
    des démances de l'age avancé. Encéphale, 1962, 4, 301-344.
Cooper, J. E., Kendall, R. E., Gurland, B. J., Sharpe, L., Copeland,
    J. R. M., and Simon, R. J. Psychiatric diagnosis in New York
    and London: A comparative study of mental hospital admissions.
    Maudsley Monograph No. 20, London: Oxford University Press,
    1972.
Cooper, J. W. Implications of drug interactions--recognition,
    incidence, and prevention. Rhode Island Medical Journal,
    1975, 58, 274-280, 287-288.
Crapper, D. R., Krishnan, S. S., and Dalton, A. J. Brain aluminum
    distribution in Alzheimer's disease and experimental neuro-
    fibrillary degeneration. Science, 1973, 180, 511-513.
Critchley, M. The neurology of psychotic speech. British Journal
    of Psychiatry, 1964, 110, 353-364.
Crook, T. Issues related to psychometric assessment of treatment
    effects in the aged. Paper presented at Geriatric Assessment
    Workshop, Los Angeles, April, 1977.
Crookes, T. G., and McDonald, K. G. Benton's visual retention
    test in the differentiation of depression and early dementia.
    British Journal of Social & Clinical Psychology, 1972, 11,
    66-69.
Cummings, J. H., Slader, G. E., and James, O. F. W. Laxative-
    induced diarrhea: A continuing clinical problem. British
    Medical Journal, 1974, 1, 537-541.
Delay, J., and Brion, E. Les demances tardives. Paris: Masson,
    1962.
Dewhurst, K., Oliver, J., Trick, K. L. K., and McKnight, A. L.
    Neuropsychiatric aspects of Huntington's disease. Contina
    Neurologia, 1969, 31, 258-268.

Dorken, H., and Greenbloom, G. C. Psychological investigations
     of senile dementia. II. The Wechsler Bellevue adult intelli-
     gence scale. Geriatrics, 1953, 8, 324-333.
Duffy, P., Wolf, J., Collins, G., DeVoe, A. G., Steeten, B., and
     Cowen, D. Possible person-to-person transmission of
     Cruetzfeldt-Jakob disease. New England Journal of Medicine,
     1974, 299, 692-693.
Earle, K. Pathology and experimental models of Huntington's
     chorea. Advances in Neurological Sciences, 1973, 1, 341-351.
Eisdorfer, C. Depression in aging. Paper presented at the VIth
     World Congress of Psychiatry, Honolulu, August, 1977.
Eisdorfer, C., and Cohen, D. Autonomic reactivity in senile
     dementia of the Alzheimer's type. In R. Katzman and R.
     Terry (Eds.), Alzheimer's disease, senile dementia and re-
     lated disorders. New York: Raven Press, in press.
Eisdorfer, C., Cohen, D., and Buckley, C. E., III. Serum immuno-
     globulins and cognition in the impaired elderly. In R.
     Katzman and R. Terry (Eds.), Alzheimer's disease, senile
     dementia and related disorders. New York: Raven Press, in
     press.
Eisdorfer, C., and Friedel, R. O. Psychotherapeutic drugs in
     aging. In M. Jarvik (Ed.), Psychopharmacology in the prac-
     tice of medicine. New York: Appleton-Century-Crofts, 1977.
Eisdorfer, C., and Stotsky, B. Intervention, treatment and re-
     habilitation of psychiatric disorders. In J. E. Birren and
     K. W. Schaie (Eds.), Handbook of the psychology of aging.
     New York: Van Nostrand Reinhold, 1977.
Fisher, C. M. Dementia and cerebral vascular disease: Dementia
     in cerebral vascular disease. In J. F. Toole, R. G. Siekert,
     and J. A. Whisnant (Eds.), Cerebral vascular diseases. New
     York: Grune and Stratton, 1968.
Folstein, M. F., Folstein, S. E., and McHugh, P. R. "Mini-mental
     state", A practical method for grading the cognitive state
     of patients for the clinician. Journal of Psychiatric Re-
     search, 1975, 12, 189-198.
Friedel, R. O., and Raskind, M. A. Psychopharmacology of aging.
     In B. Eleftheriou and M. Elias (Eds.), Special review of
     experimental aging research: Progress in biology. Palo
     Alto: E.A.R., Inc., 1977.
Gajdusek, D. C. Unconventional viruses and the origin and dis-
     appearance of Kuru. Science, 1977, 197, 943-960.
Gajdusek, D. C., and Gibbs, C. J., Jr. Familial and sporadic
     chronic neurological degenerative disorders transmitted from
     man to primates. In B. S. Meldrum and C. D. Marsden (Eds.),
     Advances in neurology, Vol. 10. New York: Raven Press,
     1975.
Gibbs, C. J., Gajdusek, D. C., Asher, D. M., Alpers, M. P., Beck,
     E., Daniel, P. M., and Matthews, W. B. Creutzfeldt-Jakob
     disease (spongioform encephalopathy): Transmission to the
     chimpanzee. Science, 1968, 161, 388-389.

Gruenberg, D. Epidemiology of senile dementia. In R. Katzman and
    R. Terry (Eds.), <u>Alzheimer's disease, senile dementia, and
    related disorders</u>. New York: Raven Press, in press.

Grunthal, E., and Wenger, O. Nachweis von erblichkeit bei der
    Alzheimer'schen krankheit nehst bemerkungen eber den alters-
    vorgans in gehirn. <u>Monatsschrift fur Psychiatrie und
    Neurologie</u>, 1939, <u>101</u>, 8-25.

Gurland, B. J. The comparative frequency of depression in various
    adult age groups. <u>Journal of Gerontology</u>, 1976, <u>31</u>, 283-292.

Gustafsen, L., and Risberg, J. Regional cerebral blood flow
    related to psychiatric symptoms in dementia with onset in
    the presenile period. <u>Acta Psychiatrica Scandinavica</u>, 1974,
    <u>50</u>, 516-538.

Hakim, S., and Adams, R. D. The special clinical problem of
    symptomatic hydrocephalus with normal cerebrospinal fluid
    hydrodynamics. <u>Journal of the Neurological Sciences</u>, 1965,
    <u>2</u>, 307-327.

Hall, P. Cyclandelate in the treatment of cerebral arteriosclero-
    sis. <u>Journal of the American Geriatric Society</u>, 1976, <u>24</u>,
    41-44.

Halstead, W. A psychometric study of senility. <u>Journal of Mental
    Science</u>, 1943, <u>89</u>, 863-873.

Hamilton, M. A rating scale for depression. <u>Journal of Neurology,
    Neurosurgery, and Psychiatry</u>, 1960, <u>23</u>, 56-61.

Helgason, T. Epidemiology of mental disorders in Iceland: A
    geriatric follow-up. Paper presented at the Fifth World
    Congress of Psychiatry, Mexico City, 1971.

Hemminki, E., and Heikkila, J. Elderly people's compliance with
    prescriptions and quality of medication. <u>Scandanavian Jour-
    nal of Social Medicine</u>, 1975, <u>3</u>, 87-92.

Holmes, T. H., and Rahe, R. H. The social readjustment rating
    scale. <u>Journal of Psychosomatic Research</u>, 1967, <u>11</u>, 213-218.

Hopkins, B., and Roth, M. Psychological test performance in
    patients over sixty. II. Paraphrenics, arteriosclerotic
    psychosis and acute confusion. <u>Journal of Mental Science</u>,
    1953, <u>99</u>, 451-463.

Huckman, M. S., Fox, J. H., and Ramsey, R. G. Computed tomography
    in the diagnosis of degenerative diseases of the brain.
    <u>Seminars in Roentgenology</u>, 1977, <u>12</u>, 63-75.

Huntington, G. On chorea. <u>Medical and Surgical Reporter</u>, 1872,
    <u>26</u>, 317-321.

Inglis, J. Memory disorder. In C. G. Costello (Ed.), <u>Symptoms
    of psychopathology</u>. New York: John Wiley, 1970.

Jacobs, J. W., Bernhard, M. R., Delgado, A., and Strain, J. J.
    Screening for organic mental syndrome in the medically ill.
    <u>Annals of Internal Medicine</u>, 1977, <u>86</u>, 40-46.

Kahn, R. L., Goldfarb, A. I., Pollack, M., and Peck, A. Brief
    objective measures for the determination of mental status
    in the aged. <u>American Journal of Psychiatry</u>, 1960, <u>117</u>,
    326-328.

Katzman, R., and Karasu, T. B.   Differential diagnosis of dementia.
    In W. S. Fields (Ed.), Neurological and sensory disorders in
    the elderly. New York: Stratton, 1975.
Katzman, R., and Pappius, H. M.   Brain electrolytes and fluid
    metabolism. Baltimore: Williams and Wilkins, 1973.
Kay, D. W. K.   Outcome and cause of death in mental disorders of
    old age: A long-term follow-up of functional and organic
    psychoses. Acta Psychiatrica Scandinavica, 1962, 38, 249–
    276.
Kay, D. W. K.   Late paraphrenia and its bearing on the etiology
    of schizophrenia. Acta Psychiatrica Scandinavica, 1963, 39,
    159–169.
Kay, D. W. K.   The epidemiology of brain deficit in the aged:
    Problems in patient identification.   In C. Eisdorfer and
    R. O. Friedel (Eds.), The cognitively and emotionally im-
    paired elderly. Chicago: Yearbook Medical Publishers, 1977.
Kay, D. W. K., Beamish, P., and Roth, M.   Old age mental disorders
    in New Castle upon Tyne, Part I: A study of prevalence.
    British Journal of Psychiatry, 1964, 110, 146–158.
Kay, D. W. K., and Roth, M.   Environmental and hereditary factors
    in the schizophrenia of old age (late paraphrenia).   Journal
    of Mental Science, 1961, 107, 649–686.
Kay, D. W. K., Roth, M., and Hopkins, B.   Affective disorders in
    the senium: (1) Their association with organic cerebral
    degeneration. Journal of Mental Science, 1955, 101, 302–318.
Kendall, R. E., Cooper, J. E., Gourlay, A. J., Copeland, J. R. M.,
    Sharpe, L., and Gurland, B. J.   Diagnostic criteria of Ameri-
    can and British psychiatrists. Archives of General Psychi-
    atry, 1972, 25, 123–130.
Kendrick, D. C.   The Kendrick battery of tests: Theoretical
    assumptions and clinical uses. British Journal of Social &
    Clinical Psychology, 1972, 40, 173–178.
Kendrick, D. C., Parboosingh, R. C., and Post, F.   A synonym
    learning test for use with elderly psychiatric subjects: A
    validation study. British Journal of Social & Clinical
    Psychology, 1965, 4, 63–71.
Kendrick, D. C., and Post, F.   Differences in cognitive status
    between healthy psychiatrically ill and diffusely brain-
    damaged elderly subjects. British Journal of Psychiatry,
    1967, 113, 75–81.
Kidd, C.   Misplacement of the elderly in the hospital. A study
    of patients admitted to geriatric and mental hospitals.
    British Medical Journal, 1962, 5313, 1491–1495.
Kirschbaum, W. R.   Jakob-Creutzfeldt disease. New York: Elsevier,
    1968.
Kirschbaum, W. R.   Jakob-Creutzfeldt disease. In J. Minkler (Ed.),
    Pathology of the nervous system. New York: McGraw-Hill,
    1971.
Kramer, M., Taube, C. A., and Redick, R. W.   Patterns of use of

psychiatric facilities by the aged: Past, present, and future. In C. Eisdorfer and M. P. Lawton (Eds.), The psychology of adult development and aging. Washington, D.C.:
American Psychological Association, 1973.

Lampert, P. W., Gajdusek, D. C., and Gibbs, C. J., Jr. Subacute
spongiform virus encephalopathies: Scrapie, kuru and
Creutzfeldt-Jakob disease. American Journal of Pathology,
1972, 68, 626-646.

Larsson, T., Sjögren, T., and Jacobson, G. Senile dementia. A
clinical sociomedical and genetic study. Acta Psychiatrica
Scandinavica, 1963, 39 (Supplement 167).

Libow, L. Senile dementia and "pseudo-senility": Clinical investigation by appropriate laboratory tests and a new mental
status evaluation technique. In C. Eisdorfer and R. O.
Friedel (Eds.), The cognitively and emotionally impaired elderly. Chicago: Yearbook Medical Publishers, 1977.

Lovett-Doust, J. W., Schneider, R. A., Talland, G. A., Walsh,
M. A., and Barker, G. B. Studies on the physiology of awareness: The correlation between intelligence and anoxemia in
senile dementia. Journal of Nervous and Mental Disease,
1953, 117, 383-398.

Lowenthal, M. J., and Berkman, P. L. Aging and mental disorder
in San Francisco. San Francisco: Jossey-Bass, 1967.

Manuelidis, E. E. Transmission of Creutzfeldt-Jakob disease
from man to guinea pig. Science, 1975, 190, 571-572.

Manuelidis, E. E., Kim, J., Angelo, J. N., and Manuelidis, L.
Serial propogation of Creutzfeldt-Jakob disease in guinea
pigs. Proceedings of the National Academy of Sciences of
the United States of America, 1976, 73, 223-227.

Marsden, C. D., and Harrison, M. J. G. Presenile dementia.
British Medical Journal, 1972, 3, 50-55.

Meynert, T. Beitrage zur differential diagnose des paralytischen
irrsinns. Wiener Medizinische Presse., 1871, 12, 645.

Miller, E. Abnormal aging. The psychology of senile and presenile dementia. New York: John Wiley, 1977.

Miller, E., and Hague, F. Some statistical characteristics of
speech in presenile dementia. Psychological Medicine, 1975,
5, 255-259.

Miller, E., and Lewis, P. Recognition memory in elderly patients
with depression and dementia: A signal detection analysis.
Journal of Abnormal Psychology, 1977, 86, 84-86.

Morrant, J. C. A. Medicines and mental illness in old age.
Canadian Psychiatric Association Journal, 1975, 20, 309-312.

Nandy, K. Immune reactions in aging brain and senile dementia.
In K. Nandy and I. Sherwin (Eds.), The aging brain and senile
dementia. New York: Plenum, 1977.

Neumann, M. A. Pick's disease. Journal of Neuropathology and
Experimental Neurology, 1949, 8, 255-282.

Neumann, M. A., and Cohn, R. Incidence of Alzheimer's disease in

a large mental hospital. Relation to senile psychosis and
psychosis with cerebral arteriosclerosis. AMA Archives of
Neurology and Psychiatry, 1953, 114, 797-811.

Newcombe, F., and Steinberg, B. Some aspects of learning and
memory function in older psychiatric patients. Journal of
Gerontology, 1964, 19, 490-493.

Newton, R. D. The identity of Alzheimer's disease and senile
dementia and their relationship to senility. Journal of
Mental Science, 1948, 94, 223-249.

Nott, R. N., and Fleminger, J. J. Presenile dementia: The difficulties of early diagnosis. Acta Psychiatrica Scandinavica,
1975, 51, 210-217.

Obrist, W. D. Electroencephalography in aging and dementia. In
R. Katzman and R. Terry (Eds.), Alzheimer's disease, senile
dementia, and related disorders. New York: Raven Press, in
press.

Olesen, J. Cerebral blood flow methods for measurement regulation:
Effects of drugs and changes in disease. Acta Neurologica
Scandinavica, 1974 (Supplement 57).

Pascarelli, E. F., and Fisher, W. Drug dependence in the elderly.
International Journal of Aging and Human Development, 1974,
5, 347-356.

Paulson, G. W., and Perrine, G., Jr. Cerebral vascular disease in
mental hospitals. In J. F. Toole, R. G. Siekert, and J. P.
Whisnant (Eds.), Cerebral vascular diseases. New York:
Grune and Stratton, 1968.

Pfeiffer, E. Functional assessment: The OARS Multidimensional
functional assessment questionnaire. Durham, N.C.: Duke
University Center for the Study of Aging and Human Development, 1975.

Post, F. The significance of affective symptoms in old age.
Maudsley Monographs No. 10. London: Oxford University
Press, 1962.

Post, F. The management and nature of depressive illness in late
life: A follow-through study. British Journal of Psychiatry,
1972, 121, 393-404.

Post, F. Dementia, depression, and pseudodementia. In D. F.
Benson and D. Blumer (Eds.), Psychiatric aspects of neurological disease. New York: Grune and Stratton, 1975.

Pratt, R. T. C. The genetics of Alzheimer's disease. In G. E.
Wolstenholme and M. O'Conner (Eds.), CIBA foundation symposium on Alzheimer's disease and related conditions. London: Churchill, 1970.

Raskind, M. Community-based evaluation and crisis intervention. A
paper presented at the Workshop on Aging, New York, October,
1976.

Raskind, M. A., Alvarez, C., Pietrzyk, M., Westerlund, K., and
Herlin, S. Helping the elderly psychiatric patient in
crisis. Geriatrics, 1976, 31, 51-56.

Raskind, M., and Eisdorfer, C. Psychopharmacology of the aged. In L. L. Simpson (Ed.), Drug treatment of mental disorders. New York: Raven Press, 1976. (a)

Raskind, M., and Eisdorfer, C. Screening for syphilis in an aged psychiatrically impaired population. Western Journal of Medicine, 1976, 125, 361-363. (b)

Redick, R. W., Kramer, M., and Taube, C. A. Epidemiology of mental illness and utilization of psychiatric facilities among older persons. In E. W. Busse and E. Pfeiffer (Eds.), Mental illness in later life. Washington, D.C.: American Psychiatric Association, 1973.

Reitan, R., and Davison, L. A. (Eds.). Clinical neuropsychology: Current status and applications. Washington, D.C.: V. H. Winston & Sons, 1974.

Resnik, H., and Kantor, J. Suicide and aging. Journal of the American Geriatric Society, 1970, 18, 152-158.

Rivera, V. M., and Meyer, J. S. Dementia and cerebrovascular disease. In J. S. Meyer (Ed.), Modern concepts of cerebrovascular disease. New York: Spectrum, 1975.

Roth, M. The natural history of mental disorder in old age. Journal of Mental Science, 1955, 101, 281-301.

Roth, M., and Morrissey, J. D. Problems in the diagnosis and classification of mental disorders in old age. Journal of Mental Science, 1952, 98, 66-80.

Roth, M., and Myers, D. H. The diagnosis of dementia. In T. Silverstone and B. Barraclough (Eds.), Contemporary psychiatry. Ashford: Headley Brothers, 1976.

Rumelhart, D. E. Introduction to human information processing. New York: John Wiley, 1977.

Salzman, C., and Shader, R. Clinical evaluation of depression in the elderly. Paper presented at Geriatric Assessment Workshop, Los Angeles, April, 1977.

Salzman, C., Kochansky, G. E., and Shader, R. Rating scales for geriatric psychopharmacology--A review. Psychopharmacology Bulletin, 1972, 8, 3-50.

Sanderson, R. E., and Inglis, J. Learning and mortality in elderly psychiatric patients. Journal of Gerontology, 1961, 16, 375-376.

Schaie, K. W., and Schaie, J. Psychological evaluation of the cognitively impaired elderly. In C. Eisdorfer and R. O. Friedel (Eds.), The cognitively and emotionally impaired elderly. Chicago: Yearbook Medical, 1977.

Schochet, S. S., Lampert, P. W., and Lindenberg, R. Fine structure of the Pick and Hirano bodies in a case of Pick's disease. Acta Neuropathologica, 1968, 11, 330-337.

Schuckit, M. A. Geriatric alcoholism and drug abuse. The Gerontologist, 1977, 17, 168-174. (a)

Schuckit, M. A. The high rate of psychiatric disorders in elderly cardiac patients. Angiology, 1977, 28, 235-247. (b)

Schuckit, M. A., Miller, P. L., and Hahlbaum, D. Unrecognized psychiatric illness in elderly medical-surgical patients. Journal of Gerontology, 1975, 30, 655-660.

Simard, D., Olesen, J., Paulsen, O. B., Lassen, N. A., and Skinhoj, E. Regional cerebral blood flow and its regulation in dementia. Brain, 1971, 94, 273-288.

Sjögren, T., Sjögren, H., and Lindgren, A. G. H. Morbus Alzheimer and morbus Pick. A genetic, clinical, and patho-anatomical study. Acta Psychiatrica et Neurologica Scandinavica, 1952, 82 (Supplement 82).

Slaby, A. E., and Wyatt, R. J. Dementia in the presenium. Illinois: Charles C Thomas, 1974.

Stengel, E. A study of the symptomatology and differential diagnosis of Alzheimer's and Pick's disease. Journal of Mental Science, 1943, 89, 1-20.

Stengel, E. Psychopathology of dementia. Proceedings of the Royal Society of Medicine, 1964, 57, 911-914.

Stern, K. Severe dementia associated with bilateral symmetrical degeneration of the thalamus. Brain, 1939, 62, 157-171.

Subby, P. A community based program for the chemically dependent elderly. Paper presented at North American Congress of Alcohol and Drug Problems, San Francisco, 1975.

Talland, G. Deranged memory: A psychonomic study of the amnesic syndrome. New York: Academic Press, 1965.

Terry, R. D., and Wisniewski, H. Ultrastructure of senile dementia and of experimental analogs. In C. Gaitz (Ed.), Aging and the brain. New York: Plenum Press, 1973.

Terry, R. D., and Wisniewski, H. Structural aspects of aging in the brain. In C. Eisdorfer and R. O. Friedel (Eds.), The cognitively and emotionally impaired elderly. Chicago: Yearbook Medical Publishers, 1977.

Tissot, R. On the nosological identity of senile dementia and of Alzheimer's disease. In C. Muller and L. Ciompi (Eds.), Senile dementia. Bern-Stuttgart: Hans Huber, 1968.

Tomlinson, B. E., Blessed, J., and Roth, M. Observations on the brains of demented old people. Journal of the Neurological Sciences, 1970, 11, 205-242.

Traub, R. D., Gajdusek, D. C., and Gibbs, C. J. Precautions in conducting biopsies and autopsies on patients with presenile dementia. Journal of Neurosurgery, 1974, 41, 394-395.

Traub, R. D., Gajdusek, D. C., and Gibbs, C. J. Precautions in autopsies on Creutzfeldt-Jakob disease. American Journal of Clinical Pathology, 1975, 64, 287.

Traub, R., Gajdusek, D. C., and Gibbs, C. J. Transmissible virus dementia: The relation of transmissible encephalopathy to Creutzfeldt-Jakob disease. In W. Lynn Smith and M. Kinsbourne (Eds.), Aging and dementia. New York: Spectrum, 1977.

van Bogaert, L. Aspects cliniques et pathologiques des atrophies pallidales et pallido-luysiennes progressives. _Journal of Neurology, Neurosurgery and Psychiatry_, 1946, _9_, 125-157.

van Bogaert, L., and Bertrand, I. Une variété d'atrophie olivo-pontine à évolution subaiguë avec troubles dementiels. _Revue Neurologique_, 1929, _36_, 165-178.

van Mansvelt, J. _Pick's disease: A syndrome of lobar cerebral atrophy_. Enschede: Van Der Loeff, 1954.

Verwoerdt, A. _Clinical geropsychiatry_. Baltimore: Williams and Wilkins, 1976.

Victor, M., Adams, R. D., and Collins, G. H. _The Wernicke-Korsakoff syndrome_. Philadelphia: Davis, 1971.

Walton, D., and Black, D. A. The validity of a psychological test of brain damage. _British Journal of Medical Psychology_, 1957, _20_, 270-279.

Walton, D., White, J. G., Black, D. A., and Young, A. J. The modified word learning test--a cross validation study. _British Journal of Medical Psychology_, 1959, _22_, 213-220.

Weingartner, H., Caine, E. D., and Ebert, M. H. Imagery, encoding and the retrieval of information from memory: Some specific encoding-retrieval changes in Huntington's disease. _Journal of Abnormal Psychology_, in press.

Winkelman, N. W. Progressive pallidal degeneration: A new clinicopathologic syndrome. _Archives of Neurology and Psychiatry_, 1932, _27_, 1-21.

Winokur, G., and Clayton, P. J. Family history studies: 1. Two types of affective disorders separated according to genetic and clinical factors. In J. Wortis (Ed.), _Recent advances in biological psychiatry_, Vol. 9. New York: Plenum Press, 1967.

Winokur, G., Clayton, P. J., and Reich, T. _Manic-depressive illness_. St. Louis: C. V. Mosby, 1969.

Wisniewski, H. M., Coblenz, J. M., and Terry, R. D. Pick's disease. A clinical and ultrastructural study. _Archives of Neurology_, 1972, _26_, 97-108.

Zung, W. W. K. A self-rating depression scale. _Archives of General Psychiatry_, 1965, _12_, 63-70.

ASSESSMENT OF ALTERED BRAIN FUNCTION IN THE AGED

Robert L. Kahn

University of Chicago

Nancy E. Miller

National Institute of Mental Health

Numerous behavioral procedures have been described in the literature which are intended to evaluate the presence and/or degree of altered brain function. Determination of the efficacy of these procedures is difficult enough in general, but becomes considerably more complicated when applied to the aged in particular. The purpose of this chapter is to consider some of the clinical issues including: the meaning of "organic brain dysfunction" that the behavioral procedures are supposed to measure, the problems of false positives and negatives, the circumstances under which the clinical examination is likely to occur, and the common problems of differential diagnosis encountered. There will be a review of the psychometric and clinical evaluation procedures commonly employed and, finally, measures which are most likely to be clinically useful will be recommended and described in detail.

CLINICAL CONCEPTS OF "ORGANIC IMPAIRMENT"

Physiological Variables

The most obvious problem is that there is no simple clinical definition of what is meant by "organic brain dysfunction." As pointed out by Reitan (1962) and by Benton and Van Allen (1972) "brain damage" is hardly a meaningful entity because of the diverse factors that enter into consideration with varying consequences for behavioral change. Some of the physiological variables include

such elements as whether the pathology is acute or chronic, deep-
seated or superficial, and diffuse or focal, let alone the lateral-
ity or other aspects of localization. Both quantitative and qualitative differences are affected by whether the damage is small or
large or whether it is due to structural change or merely a dysfunctional process accompanying a toxic or infectious condition.
According to Hebb (1949), a mere absence of brain tissue, as following surgical removal, may not produce detectable alterations in
behavior, while the presence of an irritating focus, such as scar
tissue, can lead to significant impairment.

Behavioral Differences and Similarities in Normal and Pathological Aging

The definition of "organic impairment" in the aged is an even
more difficult problem because of the controversies concerning the
nature of intellectual change with age and the relationship of
these changes to pathology of the central nervous system. The
basic issue is whether dementia can be regarded as a qualitatively
distinct entity or a quantitative extension of normal senescence.
One view is that there is a distinction between the behavioral
consequences of normal physiological changes related to aging and
those occurring with pathological conditions. Jarvik (1975) has
suggested that it is nearness to death rather than distance from
birth that is the key to impaired performance in later life. The
concept of normal physiological change has been applied to the decline in speed of response, the most common finding observed with
aging. Based on their longitudinal studies of twins, Blum, Clark,
and Jarvik (1973) claim that in contrast to other types of altered
brain behavior, psychomotor slowing is a general concomitant of
aging not related to brain changes. Similarly, Mathey (1976), re-
porting on the Bonn longitudinal study, describes simple reaction
time as measuring "functional age," but attributes the decline to
changes in blood supply and circulation. Surwillo (1968) postu-
lates that changes in the period of the alpha rhythm, the "master
timing mechanism," are related to cognitive slowing.

Reed and Reitan (1963), representing an opposing view, state
that the same type of change in psychological functions are characteristic of both diffuse cerebral dysfunction and aging, and
that the test performance of the aged may be partly explained as
reflecting "undetected cerebral pathology." Yet in one of their
own studies (Reed and Reitan, 1962) the results showed that an old
group of non-brain-damaged persons performed much better than a
group of young persons with brain damage. Muller and Grad (1974)
also found that diagnostic differences are of greater importance
than age. In studies by Overall and Gorham (1972) and by Goldstein and Shelly (1975) it was found that the pattern of change

was also qualitatively different in aged and brain damaged groups. Botwinick and Birren (1951) compared the differential Wechsler-Bellevue subtest performance of an institutional senile population with that of a control group matched for age, education, and race and found that the tests which showed the greatest decline due to normal aging were not necessarily the tests which showed the largest difference between the groups. Hallenbeck (1964) compared a uniform process with a multiple process theory of mental deterioration and concluded that mental impairment is not uniform and that different types of instruments are required for the specific condition being evaluated, aging or brain damage.

It does appear, however, that among persons with brain dysfunction age may add an additional modifying factor. Smith (1971) found that among stroke patients with aphasia older persons were more likely to be positive on the Face-Hand Test. Zarit and Kahn (1975) studied a similar population and observed that, with the same degree of sensory and motor impairment, older subjects showed greater mental impairment than younger ones. They also noted differences in adaptation; older persons with minimal brain damage demonstrated more depression and those with greater damage showed more denial, although denial correlated with severity at all ages.

Individual Differences

There is considerable controversy about the degree of relationship between structural damage and behavioral change in the elderly. In part, this derives from the many methodological difficulties such as variations in observation and record keeping, or distance in time from behavioral observation to autopsy. In an important review of the literature to that time, Wolfe (1959) concluded that no good correlation existed between the degree, distribution, and character of various abnormal brain changes and the age and state of normal function of the individual. Hamilton and Cowdry (1971) assert that the assumption that psychiatric diagnosis can identify cerebral atrophy is untenable but has been responsible for the therapeutic neglect of elderly patients. Other studies (Corsellis, 1962; Allison, 1962) emphasize that the extremes of senile behavioral deficit tend to be associated with severe and generalized cerebral pathology. There are obvious inconsistencies in autopsy studies. Twenty-five percent of the patients Corsellis (1962) considered free of any organic symptomatology were found after death to have moderate to severe brain changes. Blessed, Tomlinson and Roth (1968), using careful quantitative measures, compared behavior change and the incidence of senile plaques in the brain, and observed that, while age per se showed only a low, nonsignificant correlation with plaque count, the correlation of dementia score and number of senile plaques was +.63.

Hamilton and Cowdry (1971) have pointed out that while this is a correlation of respectable magnitude, it is far from a value that would permit prediction about an individual from a psychological assessment to anatomic diagnosis. Post (1968), responding to the lack of congruence of neuropathological and behavioral findings, has suggested that senile dementia, as a disease, may be a separate condition than senile intellectual deterioration.

Rothschild (1944) was impressed by the clinically normal patients in whom the degree of cerebral atrophy, incidence of amyloid plaques, and presence of neurofibrillary tangles were as marked as those found in individuals with severe dementing illness. He stressed that the low correlations found between severity of histological change and degree of intellectual impairment simply reflected the variable manner in which diverse personality types compensated for cerebral deterioration. Roth (1971), although supporting the quantitative theory of dementia, has nevertheless also noted the discrepancies between psychological and pathological measures and concluded that there are threshold effects, so that a certain degree of degenerative change can be accommodated within the "reserve capacity" of the brain.

Factors Affecting Individual Differences

Many individual differences in behavioral response to brain damage have been attributed to nonphysiological factors, which have been described in the literature as including work and education record, personality, sex, level of energy, patterns of adaptation, physical mobility, motivation and attitude, interest, and degree of test sophistication. The environmental context is clearly one of the important variables. Parsons (1965) found that residential isolation contributed to impairment and Lehmann (1972) suggests that chronic institutionalization tends to blur the distinction between organic and functional disorders, while Goldfarb (1972) found that a protective social setting, in contrast, may obscure the signs of impairment. Sensory deprivation, which can lead to organic mental behavior in normal young subjects (Baxton, Heron, and Scott, 1974), is apparently a major problem for the elderly, contributing to the manifestations of disturbed behavior (Hodkinson, 1973; O'Neill and Calhoun, 1975). Cameron (1941) studied elderly patients who experienced nocturnal delusions and found that he could bring on the conditions during the day by placing them in a darkened room. Bartlett (1951) observed visual hallucinations in an intellectually well-preserved man with cataracts. In a systematic study of behaviorally normal elderly patients undergoing cataract surgery Linn, Kahn, Coles, Cohen, Marshall, and Weinstein (1953) found that disturbed behavior could be elicited just by bandaging the eyes and restricting mobility, and

that over half their subjects showed a major psychosis following
surgery. The disturbed reactions occurred in those patients who,
prior to surgery, showed evidence of underlying brain dysfunction
by virtue of abnormal EEG's and positive results on the amytal test
for organic dysfunction (Weinstein, Kahn, Sugarman, and Linn,
1953). Improvement was noted in almost all cases following uncov-
ering of an eye or otherwise increasing the sensory input. This
study illustrates how older persons who are clinically normal may
have an underlying substrate of physiological damage which, in response to stress, will lead to decompensation and the manifestation of organic mental behavior.

Similar patterns are observed following other kinds of stress,
of which relocation is one of the most frequently cited examples
(Aldrich and Mendkoff, 1963). These stress reactions are often
reversible, so that one person with no change in the degree of
physiological dysfunction will provide markedly varying manifestations of behavioral impairment, with the time at which he is
tested determining whether he is "organic" or not.

Perhaps the most important nonphysiological factor affecting
behavior in the aged is affective status. Depression at any age
may interfere with cognitive function (Grinker, Miller, Sabshin,
Nunn, and Nunnally, 1961; Sternberg and Jarvik, 1976) and can present severe problems in diagnostic differentiation. But in the
aged, depression can appear as a full-blown organic mental syndrome, a condition which has been termed "pseudodementia" (Post,
1975), or "pseudosenility" (Davis, 1974). The behavior, although
organic in character, is psychogenic in etiology and not due to
any acute neurological disease or change in physical condition.

The Concept of Level of Effective Mental Function

It is important to realize that any psychological instrument
measures the response of a total organism. Much of the literature
on clinical assessment of brain function describes procedures in
terms of their ability to discriminate according to some specific
criterion, as though the question of brain dysfunction were a
dichotomous, all-or-none proposition. In fact, there is considerable variability in the nature of the possible pathological physiological condition, in the way different individuals react to the
same physiological condition, and in the way the same individual
reacts to the same physiological condition at different times. It
follows then that what any psychological test measures is not
"organicity" but behavior which represents the level of effective
mental function, the end result of the interaction of many factors,
of which the particular physiological condition is only one. It
also follows that no one procedure can be expected to have more

than a high percentage of "correct" diagnoses, and that two different procedures, each of which is clinically useful in its own
right, may not necessarily correlate too highly with each other.

False Positives and Negatives

This perspective has implications for the concept of false
positives and false negatives. Since we are dealing with tests
of effective mental function there will always be false negatives
in the sense that a person may have, for example, substantial
neuronal loss but achieve an adequate behavioral level. After a
stroke many persons may exhibit some sensory or motor impairment
and still show no disturbance of higher mental processes. They
may be "organic" technically, since there is brain pathology, but
clinically it may be far more important for management of chronic
disorders to deal with a person in terms of his functional capacity than his underlying physiological condition. Highly educated
and capable persons may show rather esoteric symptoms, hardly
fitting into the category of organic defined by any formal procedures or criteria. (One 71-year old patient, a prominent attor-
ney who had graduated first in his class at law school, showed
difficulty in reading speeches to the various chapters of the
social organization he had founded as his only notable symptom.)
False negatives would be a clinical problem most often in acute
brain disorders where recognition of the physiological dysfunction may be critical in leading to appropriate medical management,
but in such cases the qualitative aspects are more important than
quantitative behavioral indices. False positives are an especially
serious problem in dealing with the aged. The common stereotype of
the inevitability of senility in old age combined with the poor
physical health, poor cultural background, and depressed affect
often found in elderly clinical populations can easily lead to acceptance of a spurious organic diagnosis and the overlooking of
effective treatment possibilities.

PSYCHOMETRIC MEASURES OF BRAIN DYSFUNCTION

Concept of Selective Cognitive Impairment

The many types of measures employed in the assessment of brain
dysfunction in the aged are related to the interaction of a number
of factors. Most basic is the theoretical understanding of the
nature of behavioral changes with altered brain function and with
age and the extent to which these are seen as the same or separate.
Common to many theories of aging and tests of organic impairment
is the concept of selective cognitive impairment. Cattell (1963)

and Horn and Cattell (1966) studied age differences in intellectual function and distinguished two factors, crystallized (Gc) and fluid (Gf) intelligence. In their view crystallized intellectual functioning, exemplified by vocabulary, is based on stored experience and is not only resistant to impairment but may even improve with age. Fluid intelligence, exemplified by the Digit Symbol test, is characterized by new learning or problem-solving ability and shows impairment with age.

Among clinical populations the concept of selective impairment was first expounded by Babcock (1930), who believed that vocabulary skills remain relatively impervious to the ravages of brain dysfunction, providing the clinician with an accurate assessment of the patient's premorbid level of intelligence and thus indicating the amount of comparative cognitive deficit. Other tests based on the same principle included the Shipley Hartford-Retreat Conceptual Quotient and the Hunt-Minnesota Test for Organic Brain Damage.

However, there are also many studies that question the invulnerability of vocabulary and its clinical usefulness. In his review, Yates (1956) concluded that vocabulary, far from being a stable ability, does decline with brain damage, and some of the conflicting evidence on vocabulary in organic disorders has also been summarized by Payne (1961). In the New York Psychiatric Institute study of aging twins it was found (Jarvik and Falek, 1963; Blum et al., 1973) that any decline in vocabulary over time was part of the "critical loss" associated with impending mortality. (The other two components were a 2% loss on Digit Symbol or a 10% decline in Similarities, with any two of these signs considered indicative of significant physiological pathology.) Shapiro and Nelson (1955) compared the vocabulary scores of neurotic, manic depressive, schizophrenic and organic groups with normals and found that the psychiatric patients did poorer, with the organics having the lowest score. Ackelsberg (1944) studied 50 senile dements aged 60 to 81 and divided them into three groups according to their level of mental deterioration; vocabulary was related to the level of deterioration rather than to previous mental ability. Other reports have documented the deterioration of vocabulary performance in senile dementia compared to normal controls (True-blood, 1935; Pichot, 1955; Overall and Gorham, 1972) or in contrast to elderly depressives (Roth and Hopkins, 1953). In comparing aged patients with organic brain damage, depression and two normal control groups, Orme (1955) concluded that a decline in verbal ability may be the most fundamental characteristic of intellectual deterioration.

Halstead-Reitan Neuropsychological Battery

The theory of selective impairment is also expressed in studies based on the Halstead-Reitan Neuropsychological Battery for the evaluation of psychological changes associated with cerebral damage. Starting with the work of Halstead (1947) who developed a series of experimental procedures for measuring "biological intelligence" and formulated an "impairment index" indicative of organic dysfunction, the procedures were elaborated by Reitan (Reitan and Davison, 1974) who had also extended the application of the battery to the aged. Reitan (1975) indicates that the aged are impaired on those tests involving problem-solving or "immediate adaptive ability" but perform well on procedures tapping stored information. His most common procedure was to have judges rate a series of more than twenty of his tests on a stored information-problem-solving continuum and using the ranking to obtaining correlations with performance on the same task of young and old, normal and brain damaged groups (Reed and Reitan, 1963; Fitzhugh, Fitzhugh, and Reitan, 1964). On the basis of studies of aging and brain damage, Reitan (1967) concludes that the tests which provide the most reliable differences between groups with and without cerebral lesions are the Halstead Category Test, the Halstead Tactual Performance Test (time score and localization score), the Digit Symbol and Block Design Tests from the Wechsler-Bellevue Form I, and the Trail Making Test, part B. Reitan has studied such highly educated and relatively young subjects that the clinical relevance of his work for the aged is unclear. In one study of aging, 20% of his subjects had a doctorate (Reitan and Shipley, 1963) and in another the cut-off age between young and old was 35.5 (Fitzhugh et al., 1964). In another study contrasting young and old, the young subjects had a mean education of 16.6 years, significantly more than the 13.9 years for elderly subjects, although the difference was attributed only to aging (Reed and Reitan, 1963).

The Halstead-Reitan has been repeatedly investigated in determining the types of deficit associated with lesions in different areas of the brain, but Reitan (1967) has suggested that for the aged it would be desirable to identify those types of change accompanying diffuse cerebral involvement. Despite the breadth and complexity of the battery, several investigators have reported negative results when attempting to distinguish between functional and organic impairment (Orgel and McDonald, 1967; Watson, Thomas, Anderson, and Felling, 1968; Watson, Thomas, Felling, and Anderson, 1969; Barnes and Lucas, 1974; Vega and Parsons, 1967). Among those studies investigating the question of differential neuropsychological characteristics of brain-damaged versus schizophrenic groups the results revealed that neither statistical analyses nor clinical evaluations of the test protocols by trained

competent neuropsychologists succeeded in separating the two groups.
Watson et al. (1968) concluded that since the Halstead-Reitan battery was unable to distinguish organics from schizophrenics, it was
questionable that the tests were adequate indicators of brain dysfunction. Golden (1976) has suggested that, despite any diagnostic value, the Halstead-Reitan tests have clear drawbacks for general use by clinical psychologists since they require a degree of
specialization and effort difficult for a practitioner, are expen-
sive to administer, and are extremely time-consuming, requiring
from four to twelve hours. He therefore developed an abbreviated
battery and, testing a young adult population referred for evaluation, found that it was clinically useful in making a simple determination of the presence or absence of organic impairment.

The Wechsler Scales

     Wechsler's test of intelligence, first known as the Wechsler-
Bellevue (Wechsler, 1944) and later revised as the Wechsler Adult
Intelligence Scale (Wechsler, 1955) exemplifies the differential
score approach with the distinction between "Hold" and "Don't
Hold" subtests according to whether or not they were susceptible
to change with age, and the calculation of a "Deterioration Quotient" based on a comparison of the scores. The Hold tests, which
presumably showed no decline with age, include Information, Comprehension, Vocabulary, Object Assembly, and Picture Completion;
Don't Hold tests, presumably vulnerable to age, consist of Arithmetic, Digit Span, Similarities, Block Design, Digit Symbol, and
Picture Arrangement. Studies of elderly subjects have suggested
reclassification of the tests, with all performance tests regarded as Don't Hold (Berkowitz, 1953; Hunt, 1949; Madonick and
Solomon, 1947; Inglis, 1958). To some extent the Verbal-Perfor-
mance discrepancy itself has been used as the index of impairment
and studies of clinical populations have raised questions about
the usefulness of the Deterioration Quotient (Botwinick and Birren, 1951; Dorken and Greenbloom, 1953). Botwinick and Birren
(1951) found that subtests which significantly differentiated
between patients with organic mental disorders and a control group
were Information, Picture Completion, Comprehension, Arithmetic,
and Object Assembly. In their study the Information subtest, which
has been widely reported as showing little decline with age, actually showed the maximum differentiation between the two clinical
groups. The Digit Symbol test, on the other hand, is considered
very age sensitive but did not distinguish pathological deterioration. Digit Span was the poorest of all. Birren (1952) perceived
the Wechsler-Bellevue findings as supporting both the quantitative
and qualitative concepts of senility, since while there is an
accelerated decline in the ability to extract new information,
there is a loss of stored information as well.

Other Commonly Used Measures

The British frequently use the Raven's Progressive Matrices as a nonverbal test (Cunningham, Clayton, and Overton, 1975), with impairment based on comparison with a measure of vocabulary. Irving, Robinson, and McAdams (1970) compared functional and organic patients and found that while the Mill Hill Vocabulary Scale did not differentiate the groups, the Matrices was highly significant. Tests of paired associates learning have been applied by Inglis (1957) and by Isaacs (1962) but were not found significant by Irving et al. (1970). Other tests reported include the Graham-Kendall Memory for Designs and the Stroop Color-Word Interference Test (Jarvik, 1975). A number of studies have reported using the Wechsler Memory Scale, but its validity as a clinically useful test of memory has been questioned (Cohen, 1950; Howard, 1950; Coblentz, Mettis, Zingesser, Kasoff, Wisniewski, and Katzman, 1973). Savage (1975) has stressed the differences between intellectual level and intellectual learning. While using full or short forms of the Wechsler Adult Intelligence Scale for the determination of intellectual level, the learning aspect of intellectual functioning can be assessed by the Modified Word Learning Test and the Block Design Learning Test. Irving et al. (1970) have reported using the Synonym Learning Test, and Hemsi, Whitehead and Post (1968) have used the Digit Copying Test to differentiate diagnostic groups. Botwinick and Storandt (1973) have reported on the assessment of speed of motor performance on a simple cross-off test.

DeAjuriaguerra and Tissot (1968) have studied progressive cognitive changes in persons with senile dementia. They have found the Wechsler scales too limited in understanding the mechanism of "disorganization" and have, instead, focused on cognitive measures of temporal and spatial disorientation in studying "functional dissolution." Among the procedures utilized are the Piaget-Inhelder tests of conservation of matter, figure drawings, finger recognition tests, measures of apraxia (constructive, ideomotor and ideational), semantic paraphasias and phonetic errors, motor and psychomotor disintegration, and the disintegration of reflexes.

The Clinical Use of Formal Psychometric Tests

There are considerable differences of opinion reported in the literature on the clinical value of the formal psychometric tests, especially as applied to the aged. We believe that these differences depend on understandable variations in the questions asked, methodology, and population selection, and that comprehension of these factors has clear implications for the clinician. Perhaps the most common source of difference is in the selection of sub-

jects. Many studies report testing a limited group of organic
subjects typically covered by whether or not they were "testable,"
and then doing a matched pair comparison with a normal control
group. Using this method, many cognitive tasks have been shown
to differentiate pathological and normal. But the situation faced
by the practicing clinician is hardly comparable. As Heilbrun
(1962) has noted, a crucial test of a battery's clinical contribution is the ability to distinguish among various possible pathological conditions rather than between normal and pathologic. But
this is precisely where the formal psychometric tests have failed
as shown by Watson et al. (1968) and Lacks, Colbert, Harrow, and
Levine (1970) in studies of the Halstead-Reitan, for example. In
a review of over 120 research reports selected for stringent experimental design and methodological meticulousness spanning the
years 1960 through 1975, Heaton (1975) cites numerous studies suggesting that the four most common brief neuropsychological tests,
the Bender-Gestalt, Graham-Kendall Memory for Designs Test, the
Benton Visual Retention Test, and the Trail-Making Test of the
Halstead-Reitan battery, all have significant difficulty discriminating between brain damage patients and those with functional
psychiatric disorders. The tests seem especially inadequate in
differentiating the various types of organic psychopathology.
Benton and Van Allen (1972) have pointed out the lack of concor-
dance between test measures and quite obvious behavioral impairment. Hopkins and Roth (1953) found that patients with acute confusional disorders showed quite different behavioral responses
that those with senile dementia, doing quite well on tests that
are ordinarily considered indicative of organic impairment. Further factors that complicate the role of the formal psychometric
tests in the aged are the contaminating effects of cultural factors
(Vega and Parsons, 1967) and age itself. It may well be that even
if a test could discriminate organic behavior in a younger population, the criteria are not applicable to elderly subjects. The
concept of "Hold" and "Don't Hold" tests does not seem applicable,
as it is the tests ordinarily considered resistent to organic impairment in younger subjects which are the very ones to distinguish pathological physiological impairment in the aged (Botwinick and Birren, 1951; Orme, 1955; Jarvik, 1975). Among those working with clinical subjects the conventional psychometric instruments are often considered inadequate (Post, 1965; Katzman and
Karasu, 1975). Finally, there are numerous reports that the formal test batteries are too difficult to administer to more than a
minority of elderly populations. Fisher and Pierce (1967) found
a large number of untestables in community samples, and Klonoff and
Kennedy (1966) found that more than half of a hospitalized sample
were not testable.

CLINICAL SCALES

The inadequacy of psychometric procedures in clinical settings has led to the development of numerous clinical scales which
have the advantage of being far more relevant to the varied populations for whom assessment is important. Based upon the mental
status examination, which is a basic part of the training of
psychiatrists and neurologists, these scales most commonly include such items as tests of general information and of clinical
disorientation and are given such names as Mental Status Checklist
(Liftshitz, 1960), Dementia Rating Scale (Blessed et al., 1968),
or Psychogeriatric Assessment Schedule (Pattie and Gilleard, 1975).
Some of these scales also test cognitive impairment with procedures
such as the serial subtraction of 7s from 100, the ability to interpret proverbs, and tests of short-term memory such as recall
of words, the contents of a newspaper paragraph or story, or repeating digits forward and backward, but these have failed to be
useful in the differentiation of persons with altered brain function from those without cerebral pathology (Shapiro, Post, Loefving,
and Inglis, 1956). Although both experimental and clinical studies
have repeatedly emphasized the importance of the decline of short
term retention in normal and pathological aged, the most popularly
used test, Digit Span, has consistently failed to differentiate
clinically (Inglis, 1957; Botwinick and Birren, 1951).

The Mental Status Questionnaire

Illustrative of the clinical scales are two brief procedures
described by Kahn, Goldfarb, Pollack, and Peck (1960) which were
adapted for the objective determination of organic mental impairment in the aged, the Mental Status Questionnaire (MSQ) and the
Face-Hand Test (FHT). The development of these instruments occurred in the context of a large scale survey of institutionalized
aged in New York City for the purpose of comparing the populations
of the three major types of institutions, homes for the aged, nursing homes, and state hospitals. With the large number of persons
to be tested and the extremely heterogeneous social, physical, and
mental characteristics likely to be encountered, it was necessary
to develop instruments which were objective, brief, and easily
administered, and yet valid indicators of impaired mental function.

The Mental Status Questionnaire was derived from previous
studies of behavioral change with altered brain function in younger
subjects (Weinstein and Kahn, 1955). In these studies it was established that there were fundamentally different types of behavioral change occurring with acute, rapidly-developing diffuse
or deep-seated pathology, and those accompanying conditions which

were chronic, slowly developing, focal or superficial. In the
latter condition, there is difficulty in the use of denotative
symbols, such as names and numbers, and cognitive impairment is
manifested in specific functional areas such as memory. It is
this kind of deficit that seems to be measured by many of the
psychometric procedures. The other type of behavioral change is
characterized by the altered use of connotative or experiential
symbols, of which delusional denial of illness and disorientation
for place, time, and person are most prominent. In this condition
the person responds in terms of some personal or idiosyncratic
aspect of his experience rather than in terms of the intent of
the examiner; it is as though he were responding in a different
symbolic system, or that he is answering a different question than
the one asked. Thus, if a person is asked where he is his response may actually answer the question of where he would like to
be. This behavior occurs without necessarily having any explicit
cognitive or sensory deficit or lack of availability of the cues
necessary for a correct answer. The person disoriented for place,
for example, may persist in stating that he is home or at a hotel
and not in the hospital even though he is capable of reading the
name of the institution on the linens, knows that other people
call it by a different name, and despite strenuous efforts at cor-
recting him. The disoriented person may also show no impairment
at all on formal psychometric tests such as the Wechsler-Bellevue.

On the basis of this theoretical perspective an instrument
was developed that tapped both types of behavioral change. Consisting of 31 items, most of which were commonly employed in mental
status examination, the questions covered all the patterns of disorientation, tested awareness of illness, and included items on
memory for general information and calculation. After a pretest
of several hundred persons it was found that the questionnaire
could be effectively reduced to 10 items, half testing orientation
and half testing general memory. Although orientation is a complex process, it was decided to include five simple questions
testing orientation for place and time. The memory questions selected were also the simplest, as it was demonstrated that more
difficult questions would be subject to greater distortion of the
effects of cultural background.

Each item is scored as either right or wrong, so that a failure to answer a question is scored incorrect, and the total score
ranges from 0 to 10, based on the number of errors. Since it was
observed that some persons with questionable organic impairment
might still make one or two errors--most commonly missing the
correct day of the month or the name of the previous President--
the following rating system has been used: 0 to 2 errors--no or
mild brain dysfunction; 3 to 8 errors--moderate dysfunction; 9 to
10 errors--severe dysfunction. The test thus makes dichotomous

Table 1
The Mental Status Questionnaire (MSQ)

Orientation Questions

    1. Where are you now? (alternative or supplementary ques-
tions are, What place is this? What do you call this
place? What kind of place is this? If the place is an
institution the correct answer must include both the name
and <u>kind</u> of place.)

    2. What is the address of this place? (or, What is the name
of any street near here? Generally, the name of any
street bordering the block in which the interview takes
place is satisfactory.)

    3. What is the date today? (Day of the month.) Plus or
minus two days is considered correct.

    4. What is the date? (Month)

    5. What is the date? (Year)

General Information Questions

    6. How old are you? The exact age must be given, although
some persons give the age in "going on" terms.

    7. When were you born? (Month)

    8. When were you born? (Year)

    9. Who is President of the United States?

  10. Who was the President before him?

differentiation of presence or absence of brain dysfunction using
three errors as the cut-off point for a positive result, while the
absolute number of errors can provide an index of the relative
severity, useful in indicating changes over time or in making more
refined correlations with other data.

It is also important to note the qualitative value of the MSQ.
If responses to the first five questions show some of the typical
patterns of disorientation (Kahn, 1971) there is likely to be an
acute brain syndrome which is frequently reversible. In response
to a question testing orientation for place, a person with a
chronic brain disorder, and having difficulty with denotative
symbols, may be unable to remember where he is, have difficulty re-
calling the specific name or give one that contains some of the
phonetic elements or is unrelated as in aphasic misnaming. In
patients with a connotative symbolic alteration due to acute
brain damage, there will be certain characteristic patterns of
disorientation. Most commonly this includes misnaming and dis-

placement. A hospital, for example, will be called by a more
benign appellation or euphemism, such as a hotel, country club,
"place for rest and relaxation" or, as another patient said,
"menopause manor." Temporal or spatial displacement can be shown,
as by locating the place nearer his or her home, or even by naming
it as an institution in which the person had been in the past with
a more benign medical problem, such as having a baby. Another
form of disorientation is the phenomenon of reduplication in which
the patient says that he is in the annex of the institution, or
even in another institution, which has the same name and same
staff as the first one, but differs in some important aspect such
as taking only patients who are recuperating or suffering from
minor medical problems. The person with chronic brain disorder
will usually show great concern about his defect, depression or
"catastropic" anxiety, and will often appreciate or benefit from
help, recognizing the correct name when offered. With impairment
in use of connotative symbols, on the other hand, the patient is
typically bland or euphoric in affect and is uncorrectable. The
patient may, for example, be able to read the correct name of the
institution stamped on the sheets, or be told the correct name,
but he will rationalize away all the evidence and persist in his
disoriented response.

The Face-Hand Test

     The Face-Hand Test is a serendipitous procedure that developed
in the context of studying disorders of perception (Bender, 1952).
It has been observed that instances where single stimulation in a
given modality might fail to elicit any abnormality, impairment
could be demonstrated in marginal cases by simultaneous double
stimulation to the contralateral part of the body. Tactile impairment was especially manifested by contralateral stimulation of the
face and hand. It was then discovered that some patients, without
sensory impairment and without limiting errors to just one side of
the body, would make persistent errors on double simultaneous stimulation as part of a general condition of cerebral dysfunction,
and that the procedure was thus diagnostically useful (Fink, Green,
and Bender, 1952). The same pattern of errors in brain-damaged
adults has been found in children under six years of age (Fink and
Bender, 1953) and in adults over 65 (Green and Bender, 1953). It
was noted that while 37% of normal adult subjects between the ages
of 40 and 64 made errors on the first trial, nearly all were able
to correctly identify both stimuli within the next few trials.
Among the aged, however, there were persistent errors, with the
rate increasing with advancing age. In a recent study of persons
aged 60 and over seen in office practice Bender (1975) reported
that 38% made persistent errors.

The Face-Hand Test includes two series of 10 trials each, one with eyes closed and one with eyes open. Each series includes eight asymmetric combinations of face and hand, four contralateral and four ipsilateral, and two trials of symmetric stimuli, face-face and hand-hand, as shown in Table 2.

Table 2
The Face-Hand Test

1.	Right cheek	–	left hand
2.	Left cheek	–	right hand
3.	Right cheek	–	right hand
4.	Left cheek	–	left hand
5.	Right cheek	–	left cheek
6.	Right hand	–	left hand
7.	Right cheek	–	left hand
8.	Left cheek	–	right hand
9.	Right cheek	–	right hand
10.	Left cheek	–	left hand

The subject sits facing the examiner with his hands on his knees, and is given the following instructions: "I am going to touch you. Point to where I touch you." It is necessary that the instructions be specific on how to respond as verbal response leads to fewer errors than pointing. The subject is then touched with one or two brisk strokes on the cheek and dorsum of the hand.

The most common error is extinction, in which one of the stimuli is not perceived, almost always the hand. If this type of response is made on any of the first four trials the subject is asked, "Anywhere else?" This compels the subject to consider the possibility of more than one stimulus. The concept of twoness is also reinforced by the symmetric stimuli trials, which should be perceived correctly to indicate a valid procedure. If errors are made on trials 5 and 6 it indicates that the subject has a sensory impairment or is unable to follow instructions. The second most common error, and indicative of more severe pathology, is displacement, in which one stimulus is displaced to another part of the body, most often the hand stimulus being displaced to the opposite cheek. The most severe, and relatively rare, form of error has been termed "exosomesthesia," in which a stimulus is displaced outside the subject's body (Shapiro, Fink, and Bender, 1952).

If the subject continues to make errors with eyes closed, the procedure can be repeated with eyes open, giving the following

instructions: "Now I am going to touch you again. But this time
I want to keep your eyes open and watch carefully where I touch
you." It has been found that approximately 80 to 90 percent of
the persons making persistent errors with eyes closed will continue
to make errors with eyes open (Pollack, Kahn, and Goldfarb, 1958).
As the cut-off criterion for indicating brain dysfunction, any
error after the sixth trial with eyes closed can be considered a
positive response, although a person who also makes errors with
eyes open has a more severe disorder. Like the MSQ, the Face-
Hand Test also provides a continuum of dysfunction in addition to
a dichotomous differentiation of organic or not. In addition to
the qualitative aspects, the trial at which the last error was
made can serve as a functional index.

The Face-Hand Test lacks the clear cut theoretical base of
the MSQ, which may account for its more limited use despite its
empirical effectiveness (Irving et al., 1970). It apparently
is related to a general quality of attention (Bender, 1952; Pollack et al., 1958). It is also likely that some professionals
may be more reluctant to make physical contact with a subject.
In our experience we have found that although the MSQ and Face-
Hand Test only correlate between .50 and .60 with each other,
together they achieve an excellent diagnostic differentiation
(Kahn, Zarit, Hilbert, and Niederehe, 1975).

THE CLINICAL TASK

The determination of appropriate clinical instruments for the
diagnosis of altered brain function in the aged must be based in
the context of the concrete clinical task likely to be encountered.
This includes consideration of the setting in which the patient
is seen and the nature of the clinical questions being asked.
For research purposes limited to cooperative and testable subjects and comparing the subject either with himself as in a longitudinal study or to a matched normal control, there may be many
types of cognitive tasks that can show significant differences.
Even studies of psychopathologic groups are limited to selected
subjects. In the everyday clinic situation, however, where one
must work with unselected patients, where there are no comparative
data on a matched control or with the patient's previous performance, where one must deal not with comparison of normal with
abnormal but with differentiating various types of psychopathology, where one must deal with individual and not group data, and
where the patients with brain dysfunction are likely to have
associated physical illness (Harris, 1972), then the use of the
common psychometric procedures becomes questionable. Many of the
tests that have been reported as effective for research purposes

or for evaluating younger persons seem very limited for clinical
work with the aged. Many reports, as cited previously, have indicated that the standard psychometric tests cannot even be administered to more than a <u>minority</u> of elderly patients because of a
variety of factors such as physical condition, cultural limitations, lack of motivation, or extent of psychopathology.

In addition to the considerable problem of testability, the
standard procedures also seem to fall short when considering the
clinical questions likely to be asked of the test. As we have
indicated, "organic dysfunction" is not a matter of some explicit
criterion but represents a complex psycho-bio-social phenomenon.
Behavioral tests can only measure the level of effective mental
function, the resultant of this complex interaction. In the clinical situation we are not dealing with some abstract issue, such
as cognitive changes with aging, but with a concrete problem that
has immediate and important practical implications. The clinical
question is not whether this person is young or old, but whether
his behavioral change represents depression or altered brain function and, if organic, whether an acute or chronic brain syndrome.

The most common diagnostic problem that is presented with an
older population is the differentiation of organic brain dysfunction and depression. The problem arises because depression can
lead to some cognitive impairment and to apathy and psychomotor
retardation, characteristics also found with brain dysfunction.
In a general hospital, for example, it is common for medical
patients to be referred for psychiatric consultation because of
apparent depression. On examination they are frequently found
to have an acute organic behavioral syndrome secondary to medical
dysfunction, which may improve when the underlying medical condition is ameliorated. At other times the acute brain syndrome is
a manifestation of an organismic deterioration which is part of
a terminal state; in our experience this is likely to be true
when there is a generalized withdrawal marked by severe psychomotor retardation and failure to respond to questions. In an
outpatient setting there is often difficulty in diagnosing a patient with complaints such as memory difficulty, loss of interest,
inability to concentrate, or difficulty in performing usual tasks.
Frequently these patients turn out to be depressed, but because
of their age and poor performance on cognitive tasks they may just
be dismissed as senile and receive inadequate treatment.

It is often necessary that the diagnosis not only differentiate an organic from a functional disorder, but that acute and
chronic brain syndromes be distinguished because of the different
implications for etiology and treatment. A patient with chronic
brain syndrome may even develop an additional problem causing a
superimposed acute brain syndrome. The distinction between acute

and chronic is of the utmost importance since acute brain syndrome, or that component of the total disorder attributable to acute brain syndrome, is often reversible.

CONCLUSION

Considering all the aspects of the clinical situation, with the types of patients to be examined and with the kinds of clinical questions asked, it is our opinion, based on both the extensive literature and our own experience testing a wide variety of mentally-impaired elderly, that the clinical scales have an overall superiority to the formal psychometric tests. The clinical scales can be administered to many more patients, are less likely to show false positives with depression, and are most likely to differentiate acute from chronic brain syndromes. These three factors are so fundamental in clinical work with the aged that the advantage of these procedures is enormous.

There are many reports of effective clinical scales in the literature and the clinician may use any with which he feels comfortable or has had good experience, or he may even devise his own. The essential ingredients of most of these scales are tests of clinical orientation (e.g., for place and date) and simple questions of personal and general information. As one example of this kind of instrument, we have described the Kahn-Goldfarb Mental Status Questionnaire.

Although limited to only ten questions, it is obvious that one can add many other items for more detailed clinical information. Obviously, the number of information questions can be extended, although if they are not simple their diagnostic value becomes blurred. It may also be desirable to include more comprehensive questions in testing orientation. For example, testing orientation for place could include questions such as, "Where were you last night?", which elicits the confabulated journey so common in disorientation. One could also test orientation for person ("Who am I?" "Have you ever seen me before?"). In addition one could test for other types of behavior characteristic of acute brain syndrome, such as denial of illness ("What is your main trouble?" "Why did you come to the hospital?").

There are, of course, clinical questions that one may wish to ask for a given patient other than whether he is organic or not, or whether he has acute or chronic dysfunction. Questions of localization, which are not ordinarily such an issue with the aged, would have to be determined by specific cognitive tasks, a role that could be filled by some of the psychometric batteries. Or

there may be a need to test specific functional areas, such as
aphasia, which would require special testing. One of the most
interesting, but troublesome, behaviors seen after cardiovascular
accidents is "spatial inattention," in which the patient systematically neglects one side of space to a degree and manner tran-
scending specific sensory defects (Zarit and Kahn, 1974). This
behavior, which so often interferes with rehabilitation efforts,
can be easily observed and even quantified by simple tasks such as
drawing a person or a daisy or describing a picture with a variety
of persons and objects on the two sides.

     Finally, there are any number of brief procedures that indi-
vidual clinicians may use to augment their evaluation. We have
found the Face-Hand Test extremely useful in tandem with the
Mental Status Questionnaire. Another procedure that we have used
frequently is the spelling backwards of such simple words as CAT,
HAND, or WORLD. Persons with poor educational background will have
some trouble with this test so that it should be interpreted con-
servatively.

     Although we feel confident about the particular tests we have
used and recommended, there is still obviously considerable difference of opinion among psychologists working with clinical populations, which we believe is largely due to differences in selection
of subjects and resemblance to the problems actually faced in the
clinical situation. The more restrictive the criteria for the
selection of subjects, the more remote from the clinical task, the
more limited the generalizability of the procedure as a useful
clinical measure. Further research would be least useful to the
clinician if a comparison is made of a select group of patients
with a normal control group. More useful, because it was closer
to the clinical situation, would be a comparison of various pathological groups, although this would still be limited in value by
the degree of selectivity of subjects. Most useful in resolving
differences of opinion and in clinical contribution would be a study
of the diagnostic efficacy of the various tests on unselected patients. These can be either consecutive or randomly selected cases
in a variety of settings, including community mental health centers, emergency rooms, nursing homes, psychiatric referrals from
the medical and surgical units of a general hospital, and state
hospital referrals. It could be determined how the test procedures
compared in terms of the number and kinds of patients testable,
the nature of the diagnostic problems, and the contribution of the
behavioral findings to diagnosis and management. Among this mass
of physically sick, depressed, psychotic, often culturally-de-
prived older persons, there can be a really critical test of the
value of the various assessment techniques.

REFERENCES

Ackelsberg, S. B. Vocabulary and mental deterioration in senile
     dementia. Journal of Abnormal and Social Psychology, 1944,
     39, 393-406.
Aldrich, C. K. and Mendkoff, E. Relocation of the aged and dis-
     abled: A mortality study. Journal of American Geriatric
     Society, 1963, 11, 185-194.
Allison, R. S. The senile brain: A clinical study. Baltimore,
     Md.: Williams and Wilkins, 1962.
Babcock, H. An experiment in the measurement of mental deteriora-
     tion. Archives of Psychology, 1930, 18, 5-105.
Barnes, G. W. and Lucas, G. J. Cerebral dysfunction vs. psycho-
     genesis in the Halstead-Reitan test. Journal of Nervous
     and Mental Disease, 1974, 158, 50-60.
Bartlett, J. E. A. A case of organized visual hallucinations in
     an old man with cataracts and their relationship to the phe-
     nomena of the phantom limb. Brain, 1951, 74, 363-373.
Baxton, W. H., Heron, W., and Scott, T. H. Effects of decreased
     variation in the sensory environment. Canadian Journal of
     Psychology, 1954, 8, 70-76.
Bender, M. B. Disorders in perception. Springfield, Illinois:
     Charles C Thomas, 1952.
Bender, M. B. The incidence and type of perceptual deficiencies
     in the aged. In W. S. Fields (Ed.), Neurological and sensory
     disorders in the elderly. New York: Stratton Intercontin-
     ental, 1975.
Benton, A. L. and Van Allen, M. W. Aspects of neuropsychological
     assessment with cerebral disease. In C. M. Gaitz (Ed.),
     Aging and the brain, New York: Plenum Press, 1972.
Berkowitz, B. The Wechsler-Bellevue performance of white males
     past age 50. Journal of Gerontology, 1953, 8, 76-80.
Birren, J. E. A factorial analysis of the Wechsler-Bellevue
     Scale given to an elderly population. Journal of Consulting
     Psychology, 1952, 16, 399-405.
Blessed, G., Tomlinson, B. E., and Roth, M. The association be-
     tween quantitative measures of dementia and of senile change
     in the cerebral grey matter of elderly subjects. British
     Journal of Psychiatry, 1968, 114, 787-811.
Blum, J. E., Clark, E. T., and Jarvik, L. F. The New York State
     Psychiatric Institute study of aging twins. In L. F. Jarvik,
     C. Eisdorfer, and J. E. Blum (Eds.), Intellectual functioning
     in adults: Psychological and biological influences. New
     York: Springer, 1973.
Botwinick, J. and Birren, J. E. Differential decline in the
     Wechsler-Bellevue subtest in the senile psychoses. Journal
     of Gerontology, 1951, 6, 365-368.
Botwinick, J. and Storandt, M. Speed functions, vocabulary ability

and age. <u>Perceptual and Motor Skills</u>, 1973, <u>36</u>, 1123-1128.

Cameron, O. E. Studies in senile nocturnal delusion. <u>Psychiatric Quarterly</u>, 1941, <u>15</u>, 47-53.

Cattell, R. B. Theory of fluid and crystallized intelligence: A critical experiment. <u>Journal of Educational Psychology</u>, 1963, <u>54</u>, 1-22, 1963.

Coblentz, J. M., Mettis, S., Zingesser, L. H., Kasoff, S. S., Wiesniewski, H. M., Katzman, R. Presenile dementia--clinical aspects and evaluation of cerebrospinal fluid dynamics. <u>Archives of Neurology</u>, 1973, <u>29</u>, 299-308.

Cohen, J. Wechsler memory scale performance of psychoneurotic, organic and schizophrenic groups. <u>Journal of Consulting Psychology</u>, 1950, <u>14</u>, 371-373.

Corsellis, J. A. N. Mental illness and the aging brain: The distribution of pathological change in a mental hospital population. <u>Maudsley Monographs</u>, No. 9, London: Oxford University Press, 1962.

Cunningham, W. R., Clayton, V., Overton, W. Fluid and crystallized intelligence in young adulthood and old age. <u>Journal of Gerontology</u>, 1975, <u>30</u>, 53-55.

Davis, J. M. Psychopharmacology in the aged: Use of psychotropic drugs in geriatric patients. <u>Journal of Geriatric Psychiatry</u>, 1974, <u>7</u>, 145-159.

DeAjuriaguerra, J. and Tissot, R. Some aspects of psycho-neurological disintegration in senile dementia. In C. Muller and L. Ciompi (Eds.), <u>Senile dementia</u>. Bern: Hans Huber, 1968.

Dorken, H. and Greenbloom, G. C. Psychological investigation of senile dementia. II. The Wechsler-Bellevue adult intelligence scale. <u>Geriatrics</u>, 1953, <u>8</u>, 324-333.

Fink, M. and Bender, M. B. Perception of simultaneous tactile stimuli in normal children. <u>Neurology</u>, 1953, <u>3</u>, 27-34.

Fink, M., Green, M. A., and Bender, M. B. The face-hand test as a diagnostic sign of organic mental syndrome. <u>Neurology</u>, 1952, <u>2</u>, 48-56.

Fisher, J. and Pierce, R. C. Dimensions of intellectual functioning in the aged. <u>Journal of Gerontology</u>, 1967, <u>22</u>, 166-173.

Fitzhugh, K. B., Fitzhugh, L. C., and Reitan, R. M. Influence of age upon measures of problem solving and experimental background in subjects with longstanding cerebral dysfunction. <u>Journal of Gerontology</u>, 1964, <u>19</u>, 132-134.

Golden, C. J. The identification of brain damage by an abbreviated form of the Halstead-Reitan Neuropsychological Battery. <u>Journal of Clinical Psychology</u>, 1976, <u>32</u>, 821-826.

Goldfarb, A. I. Multidimensional treatment approaches. In C. Gaitz (Ed.), <u>Aging and the brain</u>. New York: Plenum Press, 1972.

Goldstein, G. and Shelly, C. H. Similarities and differences between psychological deficit in aging and brain damage. <u>Journal of Gerontology</u>, 1975, <u>30</u>, 448-455.

Green, M. A. and Bender, M. B. Cutaneous perception in the aged. Archives of Neurology and Psychiatry, 1953, 69, 577-581.

Grinker, R. R., Miller, J., Sabshin, M., Nunn, R., and Nunnally, J. C. The phenomena of depression. New York: Hoeher, 1961.

Hallenbeck, C. E. Evidence for a multiple process view of mental deterioration. Journal of Gerontology, 1964, 19, 357-363.

Halstead, W. C. Brain and intelligence: A quantitative study of the frontal lobes. Chicago: University of Chicago Press, 1947.

Hamilton, J. A. and Cowdry, E. V. Psychiatric aspects. In Cowdry, E. V. and Steinberg, F. U. (Eds.), The care of the geriatric patient, 4th Edition. St. Louis: Mosby, 1971.

Harris, R. The relationship between organic brain disease and physical status. In C. Gaitz (Ed.), Aging and the brain. New York: Plenum Press, 1972.

Heaton, R. K. Validity of neuropsychological evaluations in psychiatric settings. Paper presented at the meeting of the American Psychological Association, 1975.

Hebb, D. O. The organization of behavior. New York: John Wiley & Sons, 1949.

Heilbrun, A. B. Issues in the assessment of organic brain damage. Psychological Reports, 1962, 10, 511-515.

Hemsi, L. K., Whitehead, A., and Post, F. Cognitive functioning and cerebral arousal in elderly depressives and dements. Journal of Psychosomatic Research, 1968, 12, 145-156.

Hodkinson, H. M. Mental impairment in the elderly. Journal of Royal College of Physicians, London, 1973, 84, 579-582.

Hopkins, B. and Roth, M. Psychological test performance in patients over sixty. II. Paraphrenia, arteriosclerosis and acute confusion. Journal of Mental Science, 1953, 99, 451-463.

Horn, J. L. and Cattell, R. B. Age differences in primary mental abilities factors. Journal of Gerontology, 1966, 21, 210-220.

Howard, A. R. Diagnostic value of the Wechsler Memory Scale with selected groups of institutionalized patients. Journal of Consulting Psychology, 1950, 14, 376-380.

Hunt, W. L. The relative rate of decline of Wechsler-Bellevue "hold" and "don't hold" tests. Journal of Consulting Psychology, 1949, 13, 440-443.

Inglis, J. An experimental study of learning and "memory function" in elderly psychiatric patients. Journal of Mental Science, 1957, 103, 796-803.

Inglis, J. Psychological investigations of cognitive deficit in elderly psychiatric patients. Psychological Bulletin, 1958, 55, 197-214.

Irving, G., Robinson, R. A., and McAdams, W. The validity of some cognitive tests in the diagnosis of dementia. British Journal of Psychiatry, 1970, 117, 149-156.

Isaacs, B. A preliminary evaluation of paired-associated verbal
    learning in geriatric practice. *Gerontologia Clinica*, 1962,
    4, 43-55.
Jarvik, L. F. Thoughts on the psychobiology of aging. *American
    Psychologist*, 1975, 30, 576-583.
Jarvik, L. F. and Falek, A. Intellectual ability and survival in
    the aged. *Journal of Gerontology*, 1963, 18, 173-176.
Kahn, R. L. Psychological aspects of aging. In I. Rossman (Ed.),
    Clinical geriatrics. Philadelphia: J. B. Lippincott, 1971.
Kahn, R. L., Goldfarb, A. I., Pollack, M., and Peck, A. Brief
    objective measures for the determination of mental status in
    the aged. *American Journal of Psychiatry*, 1960, 117, 326-
    328.
Kahn, R. L., Zarit, S. H., Hilbert, N. M., and Niederehe, G.
    Memory complaint and impairment in the aged: The effect of
    depression and altered brain function. *Archives of General
    Psychiatry*, 1975, 32, 1569-1573.
Katzman, R. and Karasu, T. B. Differential diagnosis of dementia.
    In W. S. Field (Ed.), *Neurological and sensory disorders in
    the elderly*. New York: Stratton Intercontinental, 1975.
Klonoff, H. and Kennedy, M. A comparative study of cognitive func-
    tioning in old age. *Journal of Gerontology*, 1966, 21, 239-
    243.
Lacks, P. B., Colbert, J., Harrow, M., and Levine, J. Further
    evidence concerning the diagnostic accuracy of the Halstead
    Organic Test Battery. *Journal of Clinical Psychology*, 1970,
    26, 480-481.
Lehmann, H. E. Psychopharmacological aspects of geriatric medicine.
    In C. Gaitz (Ed.), *Aging and the brain*. New York: Plenum
    Press, 1972.
Lifshitz, K. Problem in the quantitative evaluation of patients
    with psychoses of the senium. *Journal of Psychology*, 1960,
    49, 295-303.
Linn, L., Kahn, R. L., Coles, R., Cohen, J., Marshall, D., and
    Weinstein, E. A. Behavior disturbances following cataract
    extraction. *American Journal of Psychiatry*, 1953, 110,
    281-289.
Madonick, M. J. and Solomon, M. The Wechsler-Bellevue Scale in
    individuals past sixty. *Geriatrics*, 1947, 2, 34-40.
Mathey, F. J. Psychomotor performance and reaction speed in old
    age. In H. Thomae (Ed.), *Patterns of aging*. Basel: S.
    Karger, 1976.
Muller, H. F. and Grad, B. Clinical-psychological, electroen-
    cephalographic and adrenocortical relationships in elderly
    psychiatric patients. *Journal of Gerontology*, 1974, 29,
    28-38.
O'Neill, P. M. and Calhoun, K. S. Sensory deficits and behavioral
    deterioration in senescence. *Journal of Abnormal Psychology*,
    1975, 84, 579-594.

Orgel, S. A. and McDonald, R. D. An evaluation of the Trail Making Test. Journal of Consulting Psychology, 1967, 31, 77–79.

Orme, J. E. Intellectual and Rorschach test performance of a group of senile dementia patients and a group of elderly depressives. Journal of Mental Science, 1955, 101, 863–870.

Overall, J. E. and Gorham, D. R. Organicity versus old age in objective and projective test performance. Journal of Consulting and Clinical Psychology, 1972, 39, 98–105.

Parsons, P. L. Mental health of Swansea's old folk. British Journal of Preventive and Social Medicine, 1965, 19, 43–47.

Pattie, A. H. and Gilleard, C. J. A brief psychogeriatric assessment schedule: Validation against psychiatric diagnosis and discharge from hospital. British Journal of Psychiatry, 1975, 27, 489–493.

Payne, R. W. Cognitive abnormalities. In H. J. Eysenck (Ed.), Handbook of abnormal psychology. New York: Basic Books, 1961.

Pichot, P. Language disturbances in cerebral disease. Archives of Neurology and Psychiatry, 1955, 74, 92–95.

Pollack, M., Kahn, R. L., and Goldfarb, A. I. Factors related to individual differences in perception in institutionalized aged. Journal of Gerontology, 1958, 13, 192–197.

Post, F. The clinical psychiatry of late life. Oxford: Pergamon Press, 1965.

Post, F. The development and progress of senile dementia in relationship to the functional psychiatric disorders of later life. In C. Muller and L. Ciompi (Eds.), Senile dementia. Bern: Hans Huber, 1968.

Post, F. Dementia, depression and pseudodementia. In D. F. Benson and D. Blumer (Eds.), Psychiatric aspects of neurologic disease. New York: Grune and Stratton, 1975.

Reed, H. B. C. and Reitan, R. M. The significance of age in the performance of a complex psychomotor task by brain-damaged and non-brain-damaged subjects. Journal of Gerontology, 1962, 17, 193–196.

Reed, H. B. C. and Reitan, R. M. A comparison of the effects of the normal aging process with the effects of organic brain-damage on adaptive abilities. Journal of Gerontology, 1963, 18, 177–179.

Reitan, R. M. Psychological deficit. In P. R. Farnsworth, O. McNemar, and Q. McNemar (Eds.), Annual review of psychology, Vol. 13. Palo Alto, California: Annual Reviews, 1962.

Reitan, R. M. Psychologic changes associated with aging and with cerebral damage. Mayo Clinic Proceedings, 1967, 43, 653–673.

Reitan, R. M. Assessment of brain-behavior relationship. In P. McReynolds (Ed.), Advances in psychological assessment, Vol. 3. San Francisco: Jossey-Bass, 1975.

Reitan, R. M. and Davison, L. A. (Eds.). Clinical neuropsychology: Current status and applications. Washington: Winston, 1974.

Reitan, R. M. and Shipley, R. E. The relationship of serum choles-
    terol changes on psychological abilities. Journal of Geron-
    tology, 1963, 18, 350-356.

Roth, M. Classification and aetiology in mental disorders of old
    age: Some recent developments. In D. W. K. Kay and A. Walk
    (Eds.), Recent developments in psychogeriatrics. Ashford,
    Kent, England: Headley, British Journal of Psychiatry,
    Special Publication No. 6, 1971.

Roth, M. and Hopkins, B. Psychological test performance in pa-
    tients over 60. I. Senile psychosis and the affective
    disorders of old age. Journal of Mental Science, 1953, 99,
    439-450.

Rothschild, D. The role of the premorbid personality in arterio-
    sclerosis psychoses. American Journal of Psychiatry, 1944,
    100, 501-505.

Savage, R. D. Psychometric techniques. In J. G. Howells (Ed.),
    Modern perspectives in the psychiatry of old age. New York:
    Brunner/Mazel, 1975.

Shapiro, M. B. and Nelson, E. H. An investigation of the nature
    of cognitive impairment in cooperative psychiatric patients.
    British Journal of Medical Psychological, 1955, 28, 239-256.

Shapiro, M. B., Post, F., Loefving, B., and Inglis, J. "Memory
    function" in psychiatric patients over 60: Some methodologi-
    cal and diagnostic implications. Journal of Mental Science,
    1956, 106, 233-246.

Shapiro, M. F., Fink, M., and Bender, M. B. Exosomesthesia or
    displacement of cutaneous sensation into extrapersonal space.
    Archives of Neurology and Psychiatry, 1952, 68, 481-490.

Smith, A. Objective indices of severity of chronic aphasia in
    stroke patients. Journal of Speech and Hearing Disorders,
    1971, 36, 167-207.

Sternberg, D. E. and Jarvik, M. E. Memory functions in depres-
    sion: Improvement with antidepressant medication. Archives
    of General Psychiatry, 1976, 33, 219-224.

Surwillo, W. W. Timing of behavior in senescence and the role of
    the central nervous system. In G. A. Talland (Ed.), Human
    aging and behavior. New York: Academic Press, 1968.

Trueblood, C. K. The deterioration of language in senility.
    Psychological Bulletin, 1935, 32, 735.

Vega, A. and Parsons, O. A. Cross-validation of the Halstead-
    Reitan tests for brain damage. Journal of Consulting
    Psychology, 1967, 31, 619-625.

Watson, C. G., Thomas, R. W., Anderson, D., and Felling, J.
    Differentiation of organics from schizophrenics at two
    chronicity levels by use of the Reitan-Halstead organic
    test battery. Journal of Consulting Psychology, 1968, 32,
    679-684.

Watson, C. G., Thomas, R. W., Felling, J., and Anderson, D.   Dif-
    ferentiation of organics from schizophrenics with the trail
    making, critical flicker fusion and light intensity matching
    tests. Journal of Clinical Psychology, 1969, 25, 130-133.
Wechsler, D. The measurement of adult intelligence, 3rd ed.
    Baltimore: Williams and Wilkins Co., 1944.
Wechsler, D. Manual for the Wechsler Adult Intelligence Scale.
    New York: The Psychological Corporation, 1955.
Weinstein, E. A. and Kahn, R. L. Denial of illness: Symbolic
    and physiological aspects. Springfield, Illinois: Charles
    C Thomas, 1955.
Weinstein, E. A., Kahn, R. L., Sugarman, L. A., and Linn, L.
    Diagnostic use of amobarbital sodium (amytal sodium) in
    organic brain disease. American Journal of Psychiatry, 1953,
    109, 889-894.
Wolfe, A. Clinical neuropathy in relation to the process of aging.
    In J. E. Birren (Ed.), The process of aging in the nervous
    system. Springfield, Illinois, 1959.
Yates, A. J. The use of vocabulary in the measurement of deteri-
    oration--A review. Journal of Mental Science, 1956, 102,
    409-440.
Zarit, S. H. and Kahn, R. L. Impairment and adaptation in chronic
    disabilities: Spatial inattention. Journal of Nervous and
    Mental Disease, 1974, 159, 63-72.
Zarit, S. H. and Kahn, R. L. Aging and adaptation to illness.
    Journal of Gerontology, 1975, 30, 67-72.

NEUROPSYCHOLOGICAL EVALUATION IN OLDER PERSONS

Diane Klisz

Geriatric Research, Education and Clinical Center

St. Louis Veterans Administration Hospital

"Clinical neuropsychology is concerned with developing knowledge about human brain-behavior relations, and with applying this
knowledge to clinical problems" (Davison, 1974, p. 3). Another way
to describe clinical neuropsychology would be to call it the study
of psychological effects of brain dysfunction (Reitan, 1966;
Davison, 1974). Since clinical neuropsychological tests have been
found to be valid indices of the status of the brain (Halstead,
1947; Reitan, 1955a; Schreiber, Goldman, Kleinman, Goldfader, and
Snow, 1976; Vega and Parsons, 1967), they may be particularly useful in gerontology; many age-related changes in psychological
functions have been attributed to the changes that occur in the
brain with aging (Wang, Obrist, and Busse, 1974; Welford and
Birren, 1965).

THE DEVELOPMENT OF A CLINICAL NEUROPSYCHOLOGICAL TEST BATTERY

The most widely used test battery for the assessment of behavioral deficits associated with brain dysfunction in adults is
the Halstead-Reitan Neuropsychological Test Battery for Adults.
This battery is the product of the work of Ward Halstead and his
former student, Ralph Reitan. The original battery, the Halstead
Test Battery for Adults, was developed by Halstead during the
period from 1935 to 1947 (Halstead, 1947). There were three major
stages in the development of the test battery. The first was a
period of naturalistic observation. During this period Halstead
supplemented his own observations of the behavior of brain-damaged
patients with those of the medical personnel, the patients' fami-

71

lies, and the patients themselves. After collecting and analyzing
these observations Halstead's next step was to develop a battery
of tests to assess the deficits that had been observed in the
patients. A battery was employed, instead of one test, because
observations had indicated that there was a wide variety of deficits in these patients; Halstead felt that a single test could
not tap all of these deficits. (The limitations of the single
test approach compared to a multi-test approach are discussed in
the introduction to this Section.) A factor analysis of the test
battery indicated that the tests were measuring four different
abilities. This analysis confirmed the view that the battery was
sensitive to a variety of deficits.

From the original battery of twenty-seven tests, thirteen
tests were selected and the performance of patients with brain damage verified by neurological tests was compared to that of psychiatric patients and normal controls. None of the people in the
psychiatric control nor in the normal control group had a diagnosis of a neurological disease. The ten tests which yielded the
largest significant differences between the brain-damaged and control groups comprised the final test battery. For each of the ten
tests selected for the final battery a criterion was established.
This criterion was the score at which there was the best differentiation of brain-damaged from control subjects. In order to summarize performance on the battery, the Impairment Index was formulated. The Impairment Index ranged from 0 to 1.0. Failure to
reach criterion on any test was represented by adding .1 to the
Index. Therefore, the Impairment Index identifies the proportion
of tests failed by the patient. Patients with brain damage were
found to have a significantly higher Impairment Index than controls (Halstead, 1947).

The third step in the development of the Halstead battery was
the validation study. The tests were administered to groups of
patients with neurologically verified brain damage and controls
(people with no neurological diagnosis of brain damage) to determine if these groups could be correctly identified as to their
neurological diagnostic category (brain-damaged or non-brain-damaged) by psychological test performance alone. Halstead's validation study confirmed the usefulness of his battery in making
this differentiation (Halstead, 1947).

Ralph Reitan has made two major contributions to the development of this neuropsychological test battery. One has been the
continuation of validation studies (Reitan, 1955a; Reitan, 1964;
Reitan and Fitzhugh, 1971; Wheeler, Burke, and Reitan, 1963). The
other has been the development of additional tests for the battery, the modification of some of the Halstead tests, and the
incorporation of other psychological tests, such as the Wechsler

intelligence tests and the MMPI, into the battery.

After thirty years of research on a wide variety of patient groups by a number of researchers, these neuropsychological tests have been found to be useful not only in differentiating the performance of non-brain-damaged people from that of brain-damaged people, but they have also been found useful in identifying behavioral deficits characteristic of subgroups of brain-damaged people. For example, patients with right hemisphere damage often have a different pattern of performance deficits than patients with left hemisphere damage (Parsons, Vega, and Burn, 1969; Reitan, 1955b; Reitan, 1964; Schrieber et al., 1976). Also, certain performance deficits are often displayed by patients with damage in a particular lobe of the brain or in a region within a lobe (Reitan, Note 1; Reitan, 1964). In addition, differences in performance deficits can be used to distinguish patients with brain damage of different etiologies, i.e., neoplastic, cerebrovascular, degenerative and traumatic (Reitan, Note 1; Reitan, 1964; Ross and Reitan, 1955; Russell, Neuringer and Goldstein, 1970).

DESCRIPTION OF THE HALSTEAD-REITAN NEUROPSYCHOLOGICAL TEST BATTERY

The present version of the battery as outlined by Reitan (Note 2) includes the following measures: Halstead's Test Battery for Adults, Trail Making Test for Adults, Aphasia Screening Test, Reitan-Klove Sensory-Perceptual Examination, Reitan-Klove Lateral Dominance Examination, Dynamometer Test, Wechsler Adult Intelligence Scale, and the Minnesota Multiphasic Personality Inventory. The following is a brief description of the components of the battery.[1]

The Halstead Test Battery for Adults

Although this battery originally consisted of ten tests, only seven of the tests are used at the present time (Halstead, 1947; Reitan, Note 2; Reitan and Davison, 1974).

The Category Test measures abstraction and organizational ability by means of concept identification problems using geometric configurations as stimuli (Reitan, 1967). The test uses a special slide projection apparatus and response key panel. The score is the total number of errors.

---

[1]For a more detailed description of the test battery see Reitan (Note 1; Note 2) or the Appendix in Reitan and Davison (1974).

The <u>Tactual Performance Test Total Time Component</u> measures ability to adapt to a novel problem-solving situation (Reitan, 1967). The patient is blindfolded and instructed to fit wooden blocks of common geometric shapes into a formboard. The patient performs the test first with the preferred hand, next with the nonpreferred hand, and finally with both hands. A measure of the time to complete the test is taken each time the test is performed and their sum yields the score.

The <u>Tactual Performance Test Memory Component</u> measures incidental memory. After the patient has fit the blocks into the formboard under the three response conditions, the formboard is hidden and the patient's blindfold is removed. The patient is required to draw the blocks placed in the formboard but never seen on a blank piece of paper. This score is the number of shapes correctly drawn.

The <u>Tactual Performance Test Localization Component</u> measures the ability to remember spatial relationships. When the patient draws the blocks he is also instructed to orient the drawings so that the placement of the blocks in the drawing corresponds to the placement of the blocks in the formboard. This score is the number of shapes that are drawn in their correct position.

The <u>Rhythm Test</u> measures ability to concentrate and to perceive differences between rhythmic sequences. Recorded pairs of tone sequences are played. After each sequence the patient judges whether the two sequences were the same or different. The score is the number of correct identifications.

The <u>Speech-Sounds Perception Test</u> measures ability to concentrate on verbal stimuli and to relate auditory verbal information to visual verbal information. A taped sequence of 60 nonsense words is presented. The patient selects the nonsense word presented from the four choices on the answer sheet by underlining the correct choice. The score is the total number of errors in identification of the nonsense words.

The <u>Finger Oscillation Test</u> measures index finger tapping speed. In this test a special finger tapping key is used. An average of five consecutive 10 sec finger tapping trials that are within five taps of each other is computed for the preferred hand and for the nonpreferred hand.

The <u>Impairment Index</u> is a summary score of the Halstead tests. Each test has a cut-off or criterion score based on norms established by Halstead (1947) with neurologically verified brain-damaged populations. The Impairment Index indicates the proportion of tests failed; therefore, it ranges from 0 to 1.0. Halstead found that an Impairment Index cut-off of .4 resulted in the best differentiation of brain-damaged from control subjects (Halstead, 1947).

Other Tests Used in the Halstead-Reitan Neuropsychological Test Battery

The following tests are often included in the Halstead-Reitan battery; however, they are not included in the Impairment Index. Criteria also based on neurological diagnosis have been established for differentiating brain-damaged patients from non-brain-damaged subjects on all of these tests (Reitan, Note 2).

The Reitan-Indiana Aphasia Screening Test is used to diagnose various types of language disorders, including inabilities or difficulties in naming common objects, spelling, reading, writing, calculating, speaking, comprehending spoken language, identifying body parts and differentiating left from right. The number and type of errors are recorded.

The Trail-Making Test for Adults measures the ability to scan a visual display and to organize sequential behaviors under time pressure. This test consists of two parts. On both parts the stimuli are a scattered arrangement of circled letters or numbers. In Part A the stimuli are circled numbers. The patient draws lines to connect the numbers in order. In Part B the stimuli are circles containing either numbers or letters. The patient draws lines from numbers to letters in alternating numerical and alphabetical order. The time to complete each part is recorded as well as the number of errors in drawing the lines.

The Reitan-Klove Sensory-Perceptual Examination is a series of tests of tactile, visual and auditory imperception. The examiner administers trials of unilateral stimulation to the patient in each modality. This is accomplished by the examiner lightly touching the patient's face or hand while the patient's eyes are closed, the examiner making discreet movements of his fingers while his arms are outstretched and the patient is focusing his eyes on the examiner's nose, and the examiner lightly rubbing his fingers by the patient's ear while standing behind the patient. Upon correct identification of unilateral stimulation, trials of bilateral sensory stimulation are interspersed with those of unilateral stimulation. The number of imperceptions during bilateral simultaneous stimulation is the score.

The Reitan-Klove Lateral Dominance Examination is an inventory of eye, hand and foot preference for various activities. The patient is asked questions about these lateral preferences and is

asked to demonstrate some of these preferences. The number of instances of left or right side preferences for the hands, eyes, and
feet are noted; these data are used in the interpretation of other
tests such as the Finger Oscillation Test, and the Tactual Performance Test.

The Dynamometer Test is a measure of grip strength. An average based on two measures of grip strength is computed for each
hand.

The Wechsler Adult Intelligence Scale (WAIS) has been incorporated into the neuropsychological test battery to replace the
Wechsler-Bellevue (Form I) (Reitan and Davison, 1974). Many of
the scores from the WAIS (Full Scale IQ, Verbal IQ, Performance
IQ, and subtest scores) are used extensively in clinical neuropsychological evaluation. The standard 550 item Minnesota Multiphasic Personality Inventory is often administered in order to
have an index of the emotional status of the patient; also some
psychiatric symptoms have been found to be indicators of brain
damage (Meier, 1969).

PERFORMANCE OF OLDER PEOPLE ON NEUROPSYCHOLOGICAL TESTS

Performance of Normal Older People

There is evidence of changes in the structure and function of
the brain in older people. Among these changes are loss of
neurons; structural alterations in surviving neurons, i.e.,
plaque formation and loss of dendrites; functional alterations in
surviving neurons, i.e., changes in neurotransmitter release;
changes in other types of brain cells, e.g., plaque formation in
glial cells.[2] The relationship between these changes and psychological functions in the elderly is a matter of speculation and
controversy at the present time (Elias, Elias and Elias, 1977;
Terry and Wisniewski, 1972). The major question underlying all
clinical neuropsychological studies of normal older people has
been: Can some of the changes in psychological functions observed
in these people be attributed to changes that are known to occur
in brain structure and function during aging? The results of
these studies have not only affirmed this question but also have
provided insights into the nature of psychological deficits in
terms of brain functions.

Since Halstead's goal was to develop a battery of tests that

---

[2]For a review of the literature of age-related changes in the central nervous system see Bondareff (1977).

would be sensitive to the condition of the brain, it is not surprising that older subjects do not perform as well as younger subjects on this battery. In fact, in view of the evidence of neuro-
anatomical and neurophysiological changes in older people, one
might suspect the usefulness of the battery in detecting the condition of the brain if no age effects had been found. However,
the sample that Halstead (1947) used to assess the discriminating
power of his test battery was relatively young. The control subjects and brain-damaged subjects ranged in age from 14 to 50 years
and 14 to 63 years, respectively. Relatively young people also
were tested in Reitan's (1955a) cross validation study of the
Halstead battery; the average age of the control and brain-damaged
groups was only 32.0 years. At the time of the early test development and cross validation studies the relationship between age
and performance was not examined.

Realizing that Halstead's data, as well as his own, had
neglected the role of age with respect to performance on the test
battery, Reitan examined this issue in a study reported in 1955
(Reitan, 1955c). The Halstead battery was administered to 327
hospitalized patients--194 brain-damaged patients with a wide variety of neurological diagnoses and 133 controls including hospitalized and non-hospitalized subjects none of whom had a diagnosis
of brain damage. All of the subjects in this study were inter-
viewed by Reitan and judged to be sufficiently alert and cooperative to participate in the testing. The subjects ranged in age
from 15 to 64 years. No data on educational levels of the groups
was presented.

To analyze the data the groups were categorized by 5 year age
intervals. The average Impairment Index for all of the groups
over the age of 45 years was above the Halstead cut-off of .4.
In other words, people without a diagnosis of brain damage but
over the age of 45 failed to achieve satisfactory performance on
five or more of Halstead's ten tests.

Results of studies following that of Reitan (1955c) have continued to confirm these findings. Reed and Reitan (1963a) administered the test battery to four groups of subjects. The raw
scores from the tests were transformed to T scores and t-test comparisons were made as follows: (1) a control group of 40 people
under the age of 40 (36 were hospitalized for a non-neurological
problem at the time of testing) were compared to a brain-damaged
group of 40 people under the age of 40 (these groups were matched
for age and education; mean age was 28.0 years, mean years of
education was 11.8); (2) a control group of 46 people between the
ages of 40 and 49 was compared to a control group of 29 people
over the age of 50 (the mean ages of these groups were 44.7 and
55.3 and the average years of education were 16.6 and 13.9, re-

spectively). The t-ratios were ranked by magnitude of their ability
to differentiate 40 to 49 year old controls from the over 50 year
old controls and young controls from the young brain-damaged. The
correlation between the rank order distributions of the t-ratios
was significant ($\underline{r}$ = .49, $\underline{p}$ < .01) indicating that the tests that
discriminated young controls from young brain-damaged were also
ones that discriminated people aged 40 to 49 from those over 50.
Reed and Reitan concluded that some of the psychological changes
associated with aging may be partially based on neuropathological
changes that are not detectable by standard neurological examination.

To determine on which tests in the battery older people show
the most decline Reed and Reitan (1963b) presented data comparing
the performance of 40 younger (mean age 28.0; mean education 11.8)
to that of 29 older individuals (mean age 53.0; mean years education 12.4) on a neuropsychological test battery. None of the subjects had a diagnosis of brain damage. Thirty-six of the forty
younger subjects were hospitalized at the time of testing while
only nine of the older subjects were hospitalized at the time of
testing. (Although it was not stated in this report, the subjects
in this study were presumably selected from those of the Reed and
Reitan [1963a] study; data on age, education and hospitalization presented in this report were identical to that presented for the
younger group in the [1963a] study.) A graph of the data from
the (1963b) study is presented in Figure 1. The older group performed significantly poorer than did the younger group on
two of three tests from the Halstead battery that have been found
to be the best indicators of the presence of brain dysfunction—
the Category Test and the Tactual Performance Test Localization
Component (Reitan, Note 1). The Trail Making Test Part B is also
one of the best indicators of the presence of brain dysfunction
(Reitan, Note 1). There was no significant difference between the
groups on performance of Trail Making Test Part B, although the
difference was in the expected direction: older subjects performed
more poorly than did younger subjects. Significant differences
between the older and younger groups in this study were also noted
in their performance of WAIS subtests that are often indicators of
brain damage: Block Design, Digit Symbol, and Object Assembly
(Wechsler, 1958).

Scores on the Halstead battery and the Trail-Making Test were
reported by Reitan and Shipley (1963) for 156 healthy men (64 aged
40 to 65 and 92 aged 25 to 39). Education level was not reported
but it may be assumed to have been higher than average since most
of the group were management or scientific employees of a large
industrial firm; at least 20% of the sample had an M.D. or Ph.D.
All t-test comparisons of scores from these tests indicated significant differences in favor of the younger subjects.

Davies (1968) studied Trail-Making Test performance in men from the ages of 20 to 80. Besides finding evidence of a significant decline on Trail-Making Test performance in the older groups, she also found that the scores of 92% of the men in the oldest age category, 70's, fell above Reitan's cut-off point (were in the impaired range). Botwinick and Storandt (1974) reported similar findings. In this study of men and women from the ages of 20 to 80, declines in Trail Making performance were associated with increasing age.

An Interpretation of the Performance Pattern of Older People on a Neuropsychological Test Battery

The first objective of a clinical neuropsychological evaluation is to determine if there is evidence of brain dysfunction. The data from studies already presented have consistently indicated that evidence of some dysfunction on clinical neuropsychological

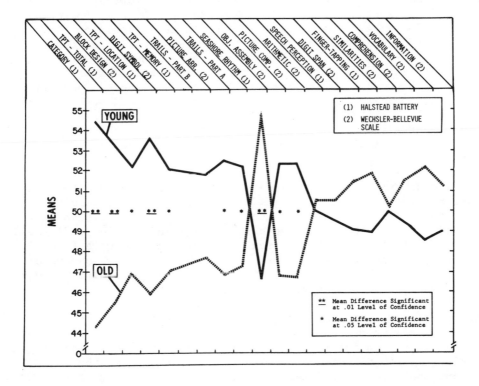

Figure 1. Performance of younger and older subjects on neuropsychological tests. (Reproduced with the permission of the Journal of Gerontology).

tests exists in normal individuals as early as age 45. The next
objective is to determine if the evidence suggests lateralized or
diffuse dysfunction. Most investigators who have addressed their
research to the communalities of performance of older people and
people with brain damage on clinical neuropsychological tests have
hypothesized that the performance of older people resembles that
of people with diffuse brain damage. To test this hypothesis discriminant function analyses of test battery measures often have
been used in order to compare test patterns from different groups.
If group differences are quantitative, one function will describe
the performance of the groups on the dependent measures. If these
differences are qualitative then more than one function will be
needed.

Overall and Gorham (1972) administered the WAIS to 299 institutionalized men between 45 and 84. There were 28 men, average
age 67.0 years, who had a diagnosis of chronic brain disease. The
271 controls who did not have a diagnosis of chronic brain disease
nor of any other neuropsychiatric disease were further categorized
by age: 27 were 45 to 54, 78 were 55 to 64, 100 were 65 to 74,
66 were 75 to 84. The multiple discriminant analysis on the subtest scores indicated that the performance of the older controls
did not resemble that of brain-damaged men. There were two significant discriminant functions: one associated with age, the
other with chronic brain disease. Analysis of subtest performance
patterns indicated that the older non-brain-damaged men performed
less well on the Similarities, Digit Symbol, Picture Arrangement
and Object Assembly subtests, relative to the Vocabulary subtest,
while the men with chronic brain syndrome suffered from a more generalized decline on all of the subtests.

To determine if psychological changes associated with aging
resemble those associated with diffuse brain damage, Goldstein
and Shelly (1975) tested 120 men between the ages of 20 and 62;
60 had a diagnosis of diffuse brain damage and 60 had no diagnosis
of brain damage. Twenty-six measures (all but two were from the
Halstead-Reitan Neuropsychological Test Battery) were obtained;
however, only four measures derived from a factor analytic study
were used in the data analysis, a multivariate analysis of variance.[3] These factor scores were labelled nonverbal memory, language ability, motor ability and psychomotor problem solving.
Based on the hypothesis of a similarity between aging and diffuse
brain damage, an interaction was predicted between age and diagnosis: little difference between the performance of the older
groups and a large difference between the performance of the
younger groups was expected. Significant main effects for age and
diagnosis were found (younger subjects performed better than older

---

[3]In the factor analytic study 455 subjects were tested.

subjects and non-brain-damaged subjects performed better than brain-damaged subjects), but the predicted interaction between age and diagnosis was not found. The relationship between age and diagnosis varied among the factor measures. On the motor ability measure the relationship was as predicted. On the language ability measure the relationship was opposite the predicted: there was little difference in performance between the younger groups and a large difference in performance between the older groups. On the nonverbal memory and psychomotor problem solving measures the differences between the younger and older groups was similar. The findings of Overall and Gorham (1972) and of Goldstein and Shelly (1975) indicate that the performance of normal older people on tests of many types of psychological abilities do not resemble that of diffusely brain-damaged people.

If neuropsychological deficits exhibited by older people do not resemble most of those exhibited by people with diffuse brain diseases, do they resemble those of people with lateralized brain damage? Schaie and Schaie (1977) noted the similarity between the performance pattern of older people on the WAIS and that found in patients with acute or chronic right hemisphere damage. As can be seen in Figure 2, there is evidence of a greater decline in the performance IQ than in the verbal IQ both in patients with right hemisphere dysfunction and in normal older people.

Some authors have suggested that the performance scale IQ decline and verbal scale IQ maintenance observed in older people are artifacts of task characteristics (Corsini and Fassett, 1953; Green, 1969; Lorge, 1936). Since older people exhibit declines in motor speed and since most of the performance tasks are timed while most of the verbal tasks are not, older people might be expected to perform more poorly than younger people on the former than on the latter tests. In a recent study Storandt (1977) administered the WAIS to groups of elderly and young subjects without using the usual time limits or bonuses. The older groups still performed significantly more poorly than did the younger groups on the performance subtests. This finding was consistent with that of an earlier investigation (Doppelt and Wallace, 1955). Thus, age-related performance scale IQ decline may not be seen as solely an artifact of the speed requirements of the tasks.

A reanalysis of the data reported by Reed and Reitan (1963a) will allow an explanation of the hypothesis that right hemisphere functions may decline more rapidly than left hemisphere functions in older people. In this study 31 tests from the Halstead-Reitan Neuropsychological Test Battery were ranked with respect to their ability to differentiate young brain-damaged patients from young controls and controls aged 40 to 49 from controls aged 50 and over. Although all of these tests have been found to distinguish the per-

formance of brain-damaged people from that of non-brain damaged
people, some of the tests have been found to be more sensitive to
dysfunction in a particular hemisphere. Based on such differences,
these tests can also be classified into three categories: those
that are relatively more sensitive to right hemisphere damage,
those that are relatively more sensitive to left hemisphere damage,
and those that are sensitive to brain damage but are not more sensitive to damage in one hemisphere than in the other one. Using
the principles outlined by Reitan (Note 1; 1964) and Reitan and
Fitzhugh (1971) the 31 tests are categorized in Tables 1 and 2 as
indicators of right hemisphere, left hemisphere, or generalized
dysfunction. As may be seen in Table 2, some of the indicators are
sensitive to both generalized and specific brain hemisphere dysfunction.

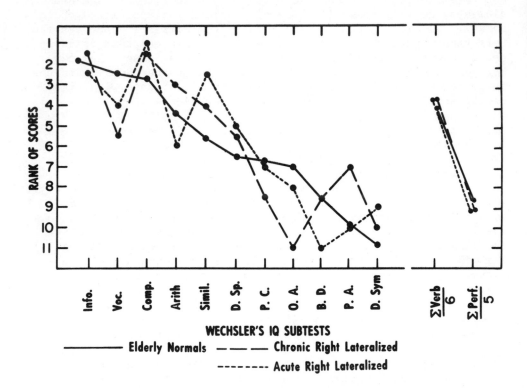

Figure 2. WAIS subtest scores of elderly normals compared to
patients with acute and chronic right hemisphere damage. (From
THE HANDBOOK OF THE PSYCHOLOGY OF AGING edited by James E. Birren
© 1977 by Litton Educational Publishing, Inc. Reprinted by permission of Van Nostrand Reinhold Company.)

Presented in Table 3 is a contingency table for the various groups used by Reed and Reitan (1963a). The question addressed in this reanalysis is which types of tests were the best (were most often ranked in the upper half) in differentiating the performance of young brain-damaged subjects from young controls and of middle-aged controls from older controls. As may be seen in Table 3 the tests which best differentiated the performance of young controls from that of young brain-damaged patients were, in order, general impairment indicators, right hemisphere impairment indicators, and left hemisphere impairment indicators. The tests which best distinguished the performance of the middle-aged controls from that of older controls were, in order, right hemisphere indicators, general impairment indicators, and, finally, left hemisphere impairment indicators. Chi-square tests revealed that right hemisphere tests were significantly better discriminators of the performance of the middle-aged controls compared to that of the older controls, $\chi^2_{(1)} = 6.0$, $\underline{p} < .02$. Although there was a tendency for right hemisphere tests to be more sensitive than the left hemisphere tests in differentiating young brain-damaged and young control groups, the chi-square test was not significant, $\chi^2_{(1)} = 3.0$, $.10 > \underline{p} > .05$. This was not surprising in view of the fact that the young brain-damaged group was comprised of a heterogenous sample of patients with respect to type of brain damage. It would appear that the conclusions of Reed and Reitan (1963a) may be modified. There is some evidence that the performance of normal older people resembles that of people with right hemisphere brain damage, suggesting that in older people there may be a greater decline of right hemisphere functions than of left hemisphere functions.

Right hemisphere functions may be more susceptible to general brain dysfunction than are left hemisphere functions. Four of the six subtests on the WAIS on which brain-damaged patients perform poorly (the Digit Span, Digit Symbol, Block Design, and Object Assembly) are performance subtests (Wechsler, 1958). In a review of studies of psychological deficits in people with diffuse brain damage Matarazzo (1972) noted that performance subtests showed a greater decline than did verbal subtests.

Semmes, Weinstein, Ghent and Teuber (1960) studied sensori-motor deficits in patients with penetrating head injuries in the left or right hemispheres. Patients with right hemisphere injuries had a greater number of sensorimotor deficits than did patients with left hemisphere injuries regardless of site of injury, i.e., inside or outside of the sensorimotor region. Based on these data and on data from subsequent studies Semmes (1968) proposed a model for hemispheric differences in the representation of various functions. According to this model, functions in the left hemisphere tend to be more focally represented while functions in the right hemisphere tend to be more diffusely organized.

Table 1

Categorization of
Wechsler-Bellevue Intelligence Scale

Test	Type of Test[a]
Performance IQ	R
Performance Weighted Score	R
Block Design	R
Digit Symbol	R
Picture Completion	R
Picture Arrangement	R
Object Assembly	R
Verbal IQ	L
Verbal Weighted Score	L
Information	L
Comprehension	L
Digit Span	L
Arithmetic	L
Similarities	L
Vocabulary	L
Full Scale IQ	G

[a]G = Test of general impairment, R = Test of right hemisphere impairment, L = Test of left hemisphere impairment

According to this model, Semmes predicted that right hemisphere functions may be more easily impaired by brain damage or dysfunction than left hemisphere functions.

Reitan and Fitzhugh (1971) studied patients who had had cerebrovascular accidents involving the left hemisphere or the right hemisphere. Group test comparisons of motor abilities indicated that significant differences were principally on left hand performance. The right hemisphere damaged group exhibited significantly poorer left hand performances than did the left hemisphere damaged group. The failure to find the opposite relationship in the right hand performance tests suggested that left hemisphere motor regions may be relatively more resistant to damage.[4]

---

[4]The testing procedure in the Tactual Performance Test can raise doubts regarding these findings. The test is performed first with the right hand and then with the left hand, thus there is a confound of order and hand use.

Jones (1971) tested verbal (left hemisphere function) and spatial (right hemisphere function) intelligence of chronic alcoholics. The performance of long term alcoholics on the spatial intelligence test was poorer than that of the nonalcoholics and short term alcoholics. No group differences were found on performance of the verbal intelligence tests. Since chronic alcoholism is known to be associated with diffuse cortical damage, Jones concluded that the results suggested a greater right hemisphere sensitivity to brain damage associated with alcoholism.

Performance of Older Brain-Damaged People

Age is a very important factor in the evaluation of deficits in brain-damaged adults. In a review of nearly three thousand cases of traumatic head injury, Russell and Smith (1961) noted that

Table 2

Categorization of Halstead-Reitan Tests

Test	Type of Test[a]
Tactual Performance Test:	
Total Time	R,G
Memory	G
Localization	R,G
Left Hand Time	R
Right Hand Time	L
Both Hands Time	G
Trail Making Test:	
Part A	R
Part B	L,G
Total Time	G
Finger Oscillation-Right Hand	L
Speech Sounds Perception	L
Rhythm	R,G
Category	G
Time-Sense Memory	G
Impairment Index	G

[a]G = Test of general impairment, R = Test of right hemisphere impairment, L = Test of left hemisphere impairment

the age of the person at the time of the injury was the single most important factor in estimating the degree of recovery. Older patients had less good prospects for recovery than did younger patients, regardless of other indices of severity of injury. A poorer prognosis for recovery and a higher mortality rate for older patients with cerebrovascular lesions has also been reported by Millikan and Moersch (1953).

Although older people have been found to exhibit more severe deficits than younger people after sustaining brain damage, opposite age trends have been reported for some symptoms. For example, epilepsy as a sequelae to traumatic head injury has been found to be negatively correlated with age (Smith, 1962). Early recovery from aphasia after a stroke (recovery which occurs spontaneously within the first month) was found to be greater in older as compared to younger aphasics (Culton, 1971). However, the aged exhibit poorer later recovery from aphasia (recovery after one month) than do the young (Eisenson, 1949; Smith, 1971).

Table 3

Number of Tests of Right Hemisphere Impairment, Left Hemisphere Impairment and General Impairment Ranking Above and Below the Median for Differentiating Young Controls from Young Brain-Damaged and Controls Aged 40 to 49 from Controls over the Age of 50[a]

Test Category	Number of Tests Above 15	Number of Tests 15 or Below
General Impairment Indicators	5 (8)	6 (3)
Right Hemisphere Impairment Indicators	9 (6)	3 (6)
Left Hemisphere Impairment Indicators	3 (2)	9 (10)

[a]The number outside the parentheses denotes the comparison of the older control groups and the number inside the parentheses denotes the comparison of the younger groups.

Reed and Reitan (1962) examined the performance of groups of subjects categorized by age and neurological status on the Tactual Performance Test, one of the most sensitive indicators of brain damage in the Halstead-Reitan battery (Reitan, Note 1). Each of the component scores from this test contribute to the Impairment Index. The mean levels of performance were, in order from best to worst: younger controls, older controls, younger brain-damaged and older brain-damaged. (Young was defined as less than 45, old as over 45; brain-damaged classification was based on neurological diagnosis). Multiple t-test comparisons were used to analyze the data. Significant differences were found between controls and brain-damaged (p < .01). No significant difference was found between young and old brain-damaged patients (.05 < p < .10). Since no age effect was observed within the brain-damaged group and since older controls performed better than did young brain-damaged patients, it can be concluded that the influence of age on a relatively sensitive measure of brain function is minimal, in comparison to the effects of brain damage.

If neuropsychological test performance is poor in older people in general, then the validity of these tests in detecting psychological concomitants of nervous system disease in older adults can be questioned. However, results of a number of studies have indicated that this concern is unfounded. In fact, deficits associated with brain damage seem to obscure those associated with age (Prigatano and Parsons, 1976; Reitan, 1955c; Reitan, 1956; Vega and Parsons, 1967). Values of the correlation coefficient between age and the Impairment Index from four studies are presented in Table 4. The correlation coefficients were always higher in the control groups than in the brain-damaged groups, this would imply that the age-related declines in neuropsychological abilities were not as great as those associated with brain damage, i.e., trauma, cerebrovascular accidents or tumors. This would be expected in view of the knowledge of the importance of rate of development of brain damage with respect to the severity of deficits associated with it. In general, rapidly developing neuropathology results in more severe psychological deficits than does slowly developing neuropathology (Joynt, 1970). Cerebrovascular accidents or rapidly growing tumors, i.e., glioblastoma multiforme, are rapidly developing neuropathologies and are usually associated with severe psychological deficits (Russell, Neuringer, and Goldstein, 1970). On the other hand, slowly growing tumors, such as meningiomas, are usually associated with mild psychological deficits (Russell, Neuringer, and Goldstein, 1970). Since the neuropathology and neuropathophysiology associated with aging seems to occur at a relatively slow rate, severe psychological deficits in normal older people would not be expected.

IMPLICATIONS FOR FUTURE RESEARCH

Clinical neuropsychological evaluation has a number of implications for future gerontological research. Clinical neuropsychological tests may be useful in the early detection and evaluation of age-related pathological processes. For example, Reitan and Shipley (1963) found evidence of impaired performance on the Halstead-Reitan Neuropsychological Test Battery in men between the ages of 40 and 65 who had elevated serum cholesterol levels; Goldman, Kleinman, Snow, Bidus and Korol (1974) observed Category Test performance impairment in middle-aged men with hypertension. Furthermore, Reitan and Shipley (1963) and Goldman, Kleinman, Snow, Bidus and Korol (1975) reported that improvement on physical mea-

Table 4

Correlation between Age and
Impairment Index in Brain-Damaged and Control Groups

Study	Age Range	Brain-Damaged r	Control r
Reitan (1955c)	15-64	.23	.54
Reitan (1956)	20-65	.37	.60
Vega and Parsons (1967)	15-74[a]	.33[b]	.57[b]
Prigatano and Parsons (1976)[c]	16-61	.44	.64

[a]Age range information for this study was obtained from Prigatano and Parsons (1976).

[b]Vega and Parsons used a revised Impairment Index which was based on T scores rather than proportion of tests failed, therefore, the sign has been changed in this table to make the correlations comparable to the others.

[c]Prigatano and Parsons also computed correlations between age and Impairment Index with education partialed out for both their data and that of Vega and Parsons. They found little change in the correlation between age and Impairment Index with education statistically controlled.

sures (serum cholesterol and blood pressure levels) was reflected
by improvement on the neuropsychological tests.

None of the men in either of the studies described above demonstrated any signs of impairment on a standard clinical neurological
examination but they did evidence signs of impairment on the clinical neuropsychological tests. The sensitivity of clinical neuropsychological tests has been noted by Klove (1974). This sensitivity, coupled with the fact that the administration of these
tests involves no apparent physical risk to the patient, makes
them appealing for use in gerontological research since certain
neurological examinations become increasingly risky for older patients (Heimburger, personal communication cited in Reitan,
1955c).

Clinical neuropsychological evaluation may also be useful in
developing rehabilitation programs for the elderly. For example,
memory abilities, especially those involved in storing new memories tend to decline with increasing age (Craik, 1977). It is
well known that mnemonic devices may serve as aids to memory and
one of the most successful mnemonic techniques is imagery (Paivio,
1971). Comparisons of younger and older subjects, however, have
indicated that the young spontaneously use imagery more often than
do the old (Hulicka and Grossman, 1967). The age-related decrease
in the tendency to use imagery as a mnemonic may be associated
with the decline in right hemisphere functions in older adults;
imagery seems to be a right hemisphere process (Seamon and
Gazzaniga, 1973). Therefore, when planning rehabilitation programs for memory impaired elderly mnemonic techniques that rely
on right hemisphere functions perhaps should be avoided in favor
of techniques that rely on left hemisphere functions, such as
verbal associations.

A clinical neuropsychological approach also may be useful in
the rehabilitation of aphasic patients. Most aphasic patients are
middle-aged or elderly people (Smith, 1971). The higher incidence
of aphasia in these age groups is due to the higher incidence of
diseases such as cerebrovascular accidents that produce aphasic
symptoms (Chusid, 1970). It has already been noted that aphasic
symptoms are more severe in older people than in younger people.
One theory of recovery from aphasia proposes that recovery is
based on shifting from the primary left hemisphere language centers to secondary language centers in the right hemisphere
(Gazzaniga, 1972). If this is the case, and if the elderly experience declines in right hemisphere functions as a normal concomitant of increased age, then slower recovery of language functions
would be expected in older people. If there are age-related
changed in severity of aphasic symptoms age-adjusted rehabilitation programs might be more successful than those that do not take
the changes into account.

Finally, for optimal use of the Halstead-Reitan Neuropsychological Test Battery with the elderly there seems to be a need to establish age norms on the battery and, perhaps, a need for the development or modification of tests within this battery. Although neuropsychological test batteries have been criticized as insensitive in elderly populations (see Chapter 2) the data on people aged 15 to 74 presented in this chapter indicate that performance deficits on these tests are more strongly associated with brain damage than with age. There are at least two ways to approach the problem of decreased sensitivity of these neuropsychological tests in the elderly. One would be to establish norms based on performance of older brain-damaged and control populations. Another would be to modify some of the tests so that the older person would not be penalized for age-related sensory and intellectual changes. Since modified forms of the Halstead-Reitan Neuropsychological Test Battery have been successfully developed for use with children from 5 to 14 (Reitan, Note 2), it seems reasonable to assume that a modified form of this battery suitable for elderly people also could be developed.

CONCLUSIONS

Normal adults over age 45 have been found to exhibit some decline in performance on tests that are sensitive to the status of the brain. However, these declines do not appear to be as great as those exhibited by brain-damaged people. Age accounts for a substantial percentage of the variance (29 to 41%) in normals but a substantially lesser percentage of the variance (only 5 to 19%) in the brain-damaged. These findings suggest that neuropathological changes associated with aging may not have as deleterious an effect as such changes usually associated with brain damage. Presumably the slow rate of development of the neuropathological changes associated with aging, compared to the fast rate of development of the neuropathological changes associated with most types of brain damage, accounts for this discrepancy in the severity of psychological deficits observed in older people compared to brain-damaged people.

Not only are the neuropsychological test performance deficits found in older people less severe than those found in brain-damaged people, they also seem to be more specific. The performance deficits of normal older people resembled those of patients with right hemisphere damage and not those of patients with left hemisphere damage. These data suggest that in normal older people there is a tendency for right hemisphere functions to decline while left hemisphere functions are better maintained.

Results of studies using clinical neuropsychological tests suggest that such evaluation may be useful in studying age-related changes in psychological functions and imply that a neuropsychological approach may be useful in developing rehabilitation programs for the elderly. Because of the difficulty of many clinical neuropsychological tests for older people, it also was concluded that the collection of new norms and the development of alternative forms of these tests may be useful.

REFERENCE NOTES

1. Reitan, R. M. The effects of brain lesions on adaptive abilities in human beings. Unpublished manuscript, Neuropsychological Laboratory, Indiana University Medical Center, Indianapolis, Indiana, 1959.

2. Reitan, R. M. Manual for administration of neuropsychological test batteries for adults and children. Unpublished manuscript, Neuropsychological Laboratory, Indiana University Medical Center, Indianapolis, Indiana, 1959.

REFERENCES

Bondareff, W. The neural basis of aging. In J. E. Birren and K. W. Schaie (Eds.), Handbook of the psychology of aging. New York: Van Nostrand Reinhold, 1977.

Botwinick, J., and Storandt, M. Memory, related functions and age. Springfield, Illinois: Charles C Thomas, 1974.

Chusid, J. G. Correlative neuroanatomy & functional neurology. Los Altos, California: Lange Medical, 1970.

Corsini, R. J., and Fassett, K. K. Intelligence and aging. Journal of Genetic Psychology, 1953, 83, 249-264.

Craik, F. I. M. Age differences in human memory. In J. E. Birren and K. W. Schaie (Eds.), Handbook of the psychology of aging. New York: Van Nostrand Reinhold, 1977.

Culton, G. L. Reaction to age as a factor in chronic aphasia in stroke patients. Journal of Speech and Hearing Disorders, 1971, 36, 563-564.

Davies, A. D. M. Measures of mental deterioration in aging and brain damage. In S. S. Chown and K. F. Riegel (Eds.), Interdisciplinary topics in gerontology (Vol. 1), Basel: Karger, 1968.

Davison, L. A. Introduction. In R. M. Reitan and L. A. Davison (Eds.), Clinical neuropsychology: Current status and applications. Washington, D.C.: V. H. Winston & Sons, 1974.

Doppelt, J. E., and Wallace, W. L. Standardization of the Wechsler
    Adult Intelligence Scale for older persons. Journal of Abnor-
    mal and Social Psychology, 1955, 51, 312–330.
Eisenson, J. Prognostic factors related to language rehabilitation
    in aphasic patients. Journal of Speech and Hearing Disorders,
    1949, 14, 262–264.
Elias, M. F., Elias, P. K., and Elias, J. W. Basic processes in
    adult developmental psychology. St. Louis: Mosby, 1977.
Gazzaniga, M. S. One brain--two minds? American Scientist, 1972,
    60, 311–317.
Goldman, H., Kleinman, K. M., Snow, M. Y., Bidus, D. R., and Korol,
    B. Correlation of diastolic blood pressure and signs of cog-
    nitive dysfunction in essential hypertension. Diseases of
    the Nervous System, 1974, 35, 571–572.
Goldman, H., Kleinman, K. M., Snow, M. Y., Bidus, D. R. and Korol,
    B. Relationship between essential hypertension and cognitive
    functioning: Effects of biofeedback. Psychophysiology,
    1975, 12, 569–573.
Goldstein, G., and Shelly, C. H. Similarities and differences be-
    tween psychological deficit in aging and brain damage. Jour-
    nal of Gerontology, 1975, 30, 448–455.
Green, R. F. Age-intelligence relationship between ages sixteen
    and sixty-four: A rising trend. Developmental Psychology,
    1969, 1, 618–627.
Halstead, W. C. Brain and intelligence: A quantitative study of
    the frontal lobes. Chicago: University of Chicago, 1947.
Hulicka, I. M., and Grossman, J. L. Age-group comparisons for
    the use of mediators in paired-associate learning. Journal
    of Gerontology, 1967, 22, 46–51.
Joynt, R. J. Presentation 5, Section II, Language disturbances in
    cerebrovascular disease. In A. L. Benton (Ed.), Behavioral
    change in cerebrovascular disease. New York: Harper & Row,
    1970.
Jones, B. M. Verbal and spatial intelligence in short and long
    term alcoholics. The Journal of Nervous and Mental Disease,
    1971, 153, 292–297.
Klove, H. Validation studies of adult clinical neuropsychology.
    In R. M. Reitan and L. A. Davison (Eds.), Clinical neuro-
    psychology: Current status and applications. Washington,
    D.C.: V. H. Winston & Sons, 1974.
Lorge, I. The influence of test upon the nature of mental decline
    as a function of age. Journal of Educational Psychology,
    1936, 27, 100–110.
Matarazzo, J. D. Wechsler's measurement and appraisal of adult
    intelligence (5th Ed.). Baltimore: Williams & Wilkins,
    1972.
Meier, M. The regional localization hypothesis and personality
    changes associated with focal cerebral lesions and ablations.
    In J. N. Butcher (Ed.), MMPI research developments and clini-
    cal applications. New York: McGraw-Hill, 1969.

Millikan, C. H., and Moersch, F. P.   Factors that influence prog-
    nosis in acute focal cerebrovascular lesions.   Archives of
    Neurology and Psychiatry, 1953, 70, 558-562.
Overall, J. E., and Gorham, D. R.   Organicity versus old age in
    objective and projective test performance.   Journal of Con-
    sulting and Clinical Psychology, 1972, 39, 98-105.
Paivio, A.   Imagery and verbal processes.   New York: Holt, Rine-
    hart, and Winston, 1971.
Parsons, O. A., Vega, Jr., A., and Burn, J.   Different psychologi-
    cal effects of lateralized brain damage.   Journal of Consult-
    ing and Clinical Psychology, 1969, 33, 551-557.
Prigatano, G. P., and Parsons, O. A.   Relationship of age and edu-
    cation to Halstead test performance in different patient
    populations.   Journal of Consulting and Clinical Psychology,
    1976, 44, 527-533.
Reed, Jr., H. B. C., and Reitan, R. M.   The significance of age in
    the performance of a complex psychomotor task by brain-damaged
    and non-brain-damaged subjects.   Journal of Gerontology, 1962,
    17, 193-196.
Reed, Jr., H. B. C., and Reitan, R. M.   A comparison of the ef-
    fects of the normal aging process with the effects of organic
    brain damage on adaptive abilities.   Journal of Gerontology,
    1963, 18, 177-179. (a)
Reed, Jr., H. B. C., and Reitan, R. M.   Changes in psychological
    test performance associated with the normal aging process.
    Journal of Gerontology, 1963, 18, 271-274. (b)
Reitan, R. M.   An investigation of the validity of Halstead's
    measures of biological intelligence.   Archives of Neurology
    and Psychiatry, 1955, 73, 28-35. (a)
Reitan, R. M.   Certain differential effects of left and right
    cerebral lesions in human adults.   Journal of Comparative
    and Physiological Psychology, 1955, 48, 474-477. (b)
Reitan, R. M.   The distribution according to age of a psychologic
    measure dependent upon organic brain functions.   Journal of
    Gerontology, 1955, 10, 338-340. (c)
Reitan, R. M.   The relationship of the Halstead impairment index
    and the Wechsler-Bellevue total weighted score to chrono-
    logic age.   Journal of Gerontology, 1956, 11, 447.
Reitan, R. M.   Psychological deficits resulting from cerebral
    lesions in man.   In J. M. Warren and K. Akert (Eds.), The
    frontal granular cortex and behavior.   New York: McGraw-
    Hill, 1964.
Reitan, R. M.   Problems and prospects in studying the psychologi-
    cal correlates of brain lesions.   Cortex, 1966, 2, 127-154.
Reitan, R. M.   Psychologic changes associated with aging and with
    cerebral damage.   Mayo Clinic Proceedings, 1967, 42, 653-673.
Reitan, R. M., and Davison, L. A. (Eds.), Clinical neuropsychology:
    Current status and applications.   Washington, D.C.:   V. H.
    Winston & Sons, 1974.

Reitan, R. M., and Fitzhugh, K. B. Behavioral deficits in groups
    with cerebral vascular lesions. Journal of Consulting and
    Clinical Psychology, 1971, 37, 215-223.
Reitan, R. M., and Shipley, R. E. The relationship of serum
    cholesterol changes to psychological abilities. Journal of
    Gerontology, 1963, 18, 350-357.
Ross, A. J., and Reitan, R. M. Intellectual and affective functions
    in multiple sclerosis: A quantitative study. A. M. A.
    Archives of Neurology and Psychiatry, 1955, 73, 663-677.
Russell, E. W., Neuringer, C., and Goldstein, G. Assessment of
    brain damage: A neuropsychological key approach. New York:
    Wiley-Interscience, 1970.
Russell, W. R., and Smith, A. Post-traumatic amnesia in closed
    head injury. Archives of Neurology, 1961, 5, 16-29.
Schaie, K. W., and Schaie, J. P. Clinical assessment and aging.
    In J. E. Birren and K. W. Schaie (Eds.), Handbook of the
    psychology of aging. New York: Van Nostrand Reinhold, 1977.
Schreiber, D. J., Goldman, H., Kleinman, K. M., Goldfader, P. R.,
    and Snow, M. Y. The relationship between independent neuro-
    psychological and neurological detection and localization of
    cerebral impairment. The Journal of Nervous and Mental
    Disease, 1976, 162, 360-365.
Seamon, J. G., and Gazzaniga, M. S. Coding strategies and cerebral
    laterality effects. Cognitive Psychology, 1973, 5, 249-256.
Semmes, J. Hemispheric specialization: A possible clue to mech-
    anism. Neuropsychologia, 1968, 6, 11-26.
Semmes, J., Weinstein, S., Ghent, L., and Teuber, H. L. Somato-
    sensory changes after penetrating brain wounds in man.
    Cambridge, Mass.: Harvard University, 1960.
Smith, A. Ambiguities in concepts and studies of "brain damage"
    and "organicity." The Journal of Nervous and Mental Disease,
    1962, 135, 311-326.
Smith, A. Objective indices of severity of chronic aphasia in
    stroke patients. Journal of Speech and Hearing Disorders,
    1971, 36, 167-207.
Storandt, M. Age, ability level, and method of administering and
    scoring the WAIS. Journal of Gerontology, 1977, 32, 175-178.
Terry, R. D., and Wisniewski, H. M. Ultrastructure of senile
    dementia and of experimental analogs. In C. M. Gaitz (Ed.),
    Aging and the brain. New York: Plenum, 1972.
Vega, Jr., A., and Parsons, O. A. Cross-validation of the Hal-
    stead-Reitan tests for brain damage. Journal of Consulting
    Psychology, 1967, 31, 619-625.
Wang, H. S., Obrist, W. D., and Busse, E. W. Neurophysiological
    correlates of the intellectual function. In E. Palmore (Ed.),
    Normal Aging II. Durham, N.C.: Duke University, 1974.
Wechsler, D. The measurement and appraisal of adult intelligence
    (4th Ed.). Baltimore: Williams & Wilkins, 1958.
Welford, A. T., and Birren, J. E. (Eds.), Behavior aging and the

nervous system. Springfield, Illinois: Charles C Thomas, 1965.

Wheeler, L., Burke, C. J., and Reitan, R. M. An application of discriminant functions to the problem of predicting brain damage using behavioral variables. Perceptual and Motor Skills, 1963, 16, 417-440.

THE ELDERLY ALCOHOLIC: SOME DIAGNOSTIC

PROBLEMS AND CONSIDERATIONS

W. Gibson Wood[1]

Department of Psychology and
All-University Gerontology Center
Syracuse University

The purpose of this chapter is to discuss some of the problems associated with the diagnosis of alcoholism[2] in the elderly.
This is an area that has received little attention both clinically and with regard to research. Thus, there are no easy
answers for the clinician who must identify and treat the elderly
alcoholic. Thus, this chapter is designed to increase the clinician's awareness of the problem and to point to some of the specific issues associated with alcoholism in the elderly.

Immediately, the question might be asked, is alcoholism
really a serious problem in the older individual? The answer is
yes, and the problem seems to be increasing as the number of
elderly persons increases in the population (Rathbone-McCuan and
Bland, 1975). Until recently, alcoholism in the elderly has not
been considered a condition requiring special consideration. The
lack of attention can be attributed to several factors. First,
alcohol abuse occurs with the greatest frequency between the ages
of 35 and 50 (Gaitz and Baer, 1971; Zimberg, 1974). With more advanced age, abstention increases and the percentage of heavy
drinkers declines (Cahalan and Cisin, 1976). Also, inspection of
general drinking patterns reveals a decline in consumption of

[1]Preparation of this chapter was supported in part by a NIAAA Postdoctoral fellowship, AA-05021-02 awarded to the author, and a
research grant from the National Institute on Aging, AG-00473-05
awarded to Dr. Merrill Elias.

[2]Alcoholism, alcohol abuse, and problem drinking will be used in
this chapter to indicate physiological and/or behavioral impairment due to chronic consumption of alcohol.

alcohol with increasing age (Cahalan and Cisin, 1976) which would
result in fewer individuals developing a drinking problem in the
later part of the life-span. Drew (1968) has suggested that alcoholism is a "self-limiting" disease because of early death,
chronic morbidity, and spontaneous recovery associated with the
disease. Support for the self-limiting nature of alcoholism has
been provided by several studies (de Lint and Schmidt, 1976;
Pattison, Abrahams, and Baker, 1974; Waller, 1976). These studies
indicate that alcoholics die at an early age from a variety of
causes, e.g., cancer of the pharynx, tongue, larynx, esophagus, and
primary carcinoma of the liver, cardiovascular disease, motor vehicle and other types of accidents, and suicide. Other reasons that
have been suggested to explain the lack of attention given the elderly alcoholic include (1) consideration of the elderly alcoholic
as a poor risk for treatment compared to a younger alcoholic; and
(2) the majority of criteria used for identifying the alcoholic
may not be appropriate for the elderly individual (Gordon, Kirchoff,
and Philipps, 1976).

While the rate of alcoholism may decline with advancing age,
this does not preclude alcohol abuse as a problem in the elderly
population. Gaitz and Baer (1971) reported that in an unselected
group of consecutive admissions of persons 60 years of age and
older to a psychiatric facility, 44% were diagnosed as having an
alcohol abuse problem. Other studies have reported percentages
ranging from approximately 18% to 28% (Schuckit and Miller, 1976;
Simon, Epstein, and Reynolds, 1968). It should be pointed out
that these percentages were based on samples of medical or psychiatric patients and thus represent very select groups. Since
these morbidity rates were not obtained from a random sample
surveyed in the community the rates may not reflect the actual
occurrence of alcoholism in the general elderly population (Zimberg, 1974). In a recent national survey (Gordon et al., 1976)
of 225 agencies involved in the referral or treatment of individuals with drinking problems, it was found that 18% of the total
population of persons being treated for alcohol abuse were 55
years of age or older. This figure may be a closer approximation
to the number of elderly alcoholics within the alcoholic population since the agencies surveyed were a cross-section of alcohol
agencies, i.e., mental health centers, counseling and referral
agencies, the alcoholism sections of hospitals. Studies of the
general population have estimated the occurrence of alcoholism
in the 55 and over age group to be between 2% and 10% (Bailey,
Haberman, and Alksne, 1965; Siassi, Crocetti, and Spiro, 1973).
The most obvious conclusion that can be drawn from these studies
is that alcohol abuse is a real problem in individuals over 55
years of age and it would appear that this problem often goes
unrecognized and untreated (Zimberg, 1974).

EFFECTS OF ALCOHOL ON THE AGING AND AGED INDIVIDUAL

Alcohol and other drugs have more pronounced behavioral and physiological effects on the older individual as compared to the younger (see reviews: Friedel and Raskind, 1976; Omenn, 1976; Shuster, 1976; Wood, 1976a). The effects of alcohol are seen with both aged and aging laboratory animals and humans. Aging laboratory animals, i.e., rats and mice, show a greater toxicity to alcohol (Ernst, Dempster, Yee, St. Dennis, and Nakano, 1976; Wiberg, Trenholm, and Coldwell, 1970), consume less alcohol (Goodrick, 1967; 1975; Wood, 1976b), and show more behavioral impairment than do younger animals (Ernst et al., 1976). Older animals metabolize alcohol at a slower rate (Wiberg et al., 1970) and also exhibit a greater change in brain chemistry in response to alcohol when compared to younger animals (Sun, Ordy, and Samorajski, 1975).

In humans, behavior is more impaired in older alcoholics as compared to younger alcoholics on such measures as the WAIS, Halstead Neuropsychological Test Battery, and information processing tasks (Cermak and Ryback, 1976; Klisz and Parsons, 1977; Wood, Elias, Schultz, Dineen, and Pentz, 1977). This impairment cannot be accounted for by differences in age groups as comparisons of young and old nonalcoholics reveal significantly less differences between the age groups. As in the animal literature, humans in both cross-sectional and longitudinal studies appear to be more affected by alcohol physiologically than are younger subjects. Older individuals show a higher blood level of alcohol than do younger persons when administered equal doses of alcohol based on body weight (Vestal, McGuire, Tobin, Andres, Norris, and Mezey, 1977). The difference in blood level has been attributed to changes in body composition, i.e., reduced body water and decreased lean body mass (Vestal et al., 1977). Changes in brain chemistry also have been seen in human brain autopsy samples treated with alcohol from young, middle age, and aged individuals (Sun et al., 1975). These changes are more pronounced in the tissue samples of older individuals.

Clinically, it has been observed that excessive consumption of alcohol can result in pathological changes in various organ systems of the body, e.g., liver disease (Feinman and Lieber, 1974), respiratory disease (Lyons and Saltzman, 1974), gastrointestinal pathology (Lorber, Dinoso, and Chey, 1974), and diseases of the nervous system (Dreyfus, 1974) and cardiovascular system (Burch and Giles, 1974). Chronic alcoholism also can result in malnutrition (Hillman, 1974) that can exacerbate or cause some of the diseases mentioned above.

   A majority of the organ systems listed above also are systems that show a high morbidity with advancing age (Timiras,
1972). Thus, excessive drinking in an elderly individual could
very easily precipitate or exaggerate pathological changes in
these systems. Also, there is some similarity between alcohol-
induced pathology and pathological changes that frequently occur
in the elderly. This similarity could result in the primary
diagnosis of alcoholism being missed in the elderly alcoholic
(Gordon et al., 1976).

   A final word on the effects of alcohol on the aged and
aging individual concerns the interaction of various drugs with
alcohol. The combination of certain drugs with alcohol can result in a synergistic effect, leading to coma or death. Elderly
individuals take more medication than any other age group (Goldberg and Roberts, 1976) which makes them a high risk group in
regard to alcohol and drug interactions.

   The conclusion that can be drawn from this brief discussion
of the effects of alcohol on the aged individual is that excessive alcohol consumption is very serious in this age group, both
behaviorally and physiologically. Moreover, the diagnosis of
alcoholism can be confused with pathological changes that sometimes occur in older individuals. Also, these individuals may be
a high risk group in regard to the interaction of alcohol with
various drugs.

DIAGNOSIS OF CHRONIC ALCOHOLISM

Diagnosis in the General Population

   How do you identify an individual with a drinking problem?
At first glance this question appears to be answered easily. Unfortunately this is not the case. There is no sharp, objective
demarcation between the problem drinker and the person who does
not abuse alcohol. Part of the ambiguity and confusion in the
diagnosis of alcoholism is that there is no clearly defined and
objective tests used in determining this condition.[3] Barry
(1974, p. 55), in describing the problem of identifying the alcohol abuser, states that "the same person may be defined as alcoholic and nonalcoholic at different times or at the same time by
different observers." Thus, identifying the alcoholic is not

---

[3]Recent work by Shaw et al. (Science, 1976, 194, 1057-1058) has
indicated that the ratio of plasma alpha amino-n-butyric acid to
leucine may be one of the first reliable biochemical markers used
to diagnose alcoholism.

only a matter of recognizing the symptoms of the disease but the
diagnosis also can be influenced by factors not directly associated with alcohol abuse. Factors such as the sex, culture, and
age of both the patient and clinician can make the diagnosis difficult.

Despite the difficulties encountered in identifying the alcoholic, several different sets of criteria have been proposed and
used in diagnosing this condition (Criteria Committee, National
Council on Alcoholism, 1972; Fisher, Mason, and Fisher, 1976;
Jacobson, 1976; Jellinek, 1960). For the purpose of this discussion, the diagnostic criteria developed by the National Council on Alcoholism (1972) will be used to determine the appropriateness of the conventional diagnostic criteria in the identification of the elderly alcoholic. This decision was based on the
fact that these criteria include both extensive physiological
and behavioral symptomology. These criteria have been criticized for their lengthiness, emphasis on the later stages of
alcoholism, and the lack of testing for validity and reliability.
While the criteria are not without problems, they do represent
a set of diverse criteria as opposed to a single criterion and
therefore provide one of the few eclectic estimations of the
effects of alcohol on the individual.

The extensive list of symptoms (criteria) directly or indirectly associated with alcoholism is shown in Tables 1 and 2.[4]
The criteria are defined as "major" or "minor," depending on
their weight in the diagnosis. Major criteria are listed in
Table 1 and minor criteria are listed in Table 2. Both major and
minor criteria are grouped under two "Tracks." Track I lists
Physiological and Clinical criteria, and Track II includes
Behavioral, Psychological, and Attitudinal criteria. Track I
consists of several physical symptoms and clinical diseases that
are directly or indirectly associated with chronic alcohol
abuse. Track II contains criteria relating to drinking behavior,
employment history and family and social interactions. Also,
each criterion is defined by one of three "Diagnostic Levels."

---

[4]This article by the Criteria Committee, National Council on
Alcoholism, is reprinted with permission from the American
Journal of Psychiatry, Vol. 129, pp. 127-135, 1972, copyright
American Psychiatric Association, 1972; and Annals of Internal
Medicine, Vol. 17, pp. 249-258, 1972, copyright Annals of Inter-
nal Medicine 1972. Reprints of the "Criteria for the Diagnosis
of Alcoholism" are available from the National Council on Alco-
holism, 733 Third Avenue, New York, N.Y. 10017.

Diagnostic Level 1. Classical, definite, obligatory: A person who fits this criterion must be diagnosed as being alcoholic.

Diagnostic Level 2. Probable, frequent, indicative: A person who satisfies this criterion is under strong evidence of alcoholism; other corroborative evidence should be obtained.

Diagnostic Level 3. Potential, possible, incidental: These manifestations are common in people with alcoholism, but do not in themselves give a strong indication of its existence. They may arouse suspicion, but significant other evidence is needed before the diagnosis is made (National Council on Alcoholism, 1972, pp. 129-130).

In order that the diagnosis of alcoholism can be made, one or more Diagnostic Level 1 criteria, or several Diagnostic Level 2, both major or minor, should be observed. Kissin (1974, p. 30) suggests that the utilization of the criteria as well as an understanding of the clinical development of the disease "provides a rational basis for the diagnosis of chronic alcoholism."

Table 1

Major Criteria for the Diagnosis of Alcoholism

CRITERION	DIAGNOSTIC LEVEL	CRITERION	DIAGNOSTIC LEVEL
TRACK I. PHYSIOLOGICAL AND CLINICAL		2. Evidence of *tolerance* to the effects of alcohol. (There may be a decrease in previously high levels of tolerance late in the course.) Although the degree of tolerance to alcohol in no way matches the degree of tolerance to other drugs, the behavioral effects of a given amount of alcohol vary greatly between alcoholic and nonalcoholic subjects.	
A. Physiological Dependency			
1. Physiological dependence as manifested by evidence of a *withdrawal syndrome* when the intake of alcohol is interrupted or decreased without substitution of other sedation. It must be remembered that overuse of other sedative drugs can produce a similar withdrawal state, which should be differentiated from withdrawal from alcohol.			
a) Gross tremor (differentiated from other causes of tremor)	1	a) A blood alcohol level of more than 150 mg. without gross evidence of intoxication.	1
b) Hallucinosis (differentiated from schizophrenic hallucinations or other psychoses)	1	b) The consumption of one-fifth of a gallon of whiskey or an equivalent amount of wine or beer daily, for more than one day, by a 180-lb. individual	1
c) Withdrawal seizures (differentiated from epilepsy and other seizure disorders)	1	3. Alcoholic "blackout" periods. (Differential diagnosis from purely psychological fugue states and psychomotor seizures.)	2
d) Delirium tremens. Usually starts between the first and third day after withdrawal and minimally includes tremors, disorientation, and hallucinations	1	B. Clinical Major Alcohol-Associated Illnesses Alcoholism can be assumed to exist if major alcohol-associated illnesses develop in a person who drinks regularly. In such individuals, evidence of physiological and psychological dependence should be searched for.	

Diagnosis of the Elderly Alcoholic

In examining the criteria described in Tables 1 and 2 it would seem that many of the psychological and clinical symptoms and the behavioral and psychological symptoms, particularly the Diagnostic Level 2 and 3 criteria, occur frequently in elderly individuals who do not have a drinking problem. For example, conditions such as brain damage, gastrointestinal disorders, and cardiovascular disease all show an increase in frequency in older adults (Timiras, 1972). Moreover, the criteria under the Behavioral, Psychological, and Attitudinal track are highly weighed toward factors that are significantly associated with the elderly population, e.g., depressive or mood-cyclic disorders, changes in employment, economic, and marital status. The similarities between changes associated with alcohol abuse and those associated with aging suggest that the criteria may be confounded with age. Unless alcoholism is suspected, an elderly individual showing some of the symptoms listed in Tables 1 and 2 might be viewed as a person with both physical and psychological problems associated with aging alone. Thus, the probability of alcoholism being cor-

Table 1 (Cont'd)

CRITERION	DIAGNOSTIC LEVEL	CRITERION	DIAGNOSTIC LEVEL
Fatty degeneration in absence of other known cause	2		
Alcoholic hepatitis	1	TRACK II. BEHAVIORAL, PSYCHOLOGICAL AND ATTITUDINAL	
Laennec's cirrhosis	2		
Pancreatitis in the absence of cholelithiasis	2	All chronic conditions of psychological dependence occur in dynamic equilibrium	
Chronic gastritis	3	with intrapsychic and interpersonal consequences. In alcoholism, similarly, there are	
Hematological disorders:		varied effects on character and family. Like	
Anemia: hypochromic, normocytic, macrocytic, hemolytic with stomato-		other chronic relapsing diseases, alcoholism produces vocational, social, and physical	
cytosis, low folic acid	3	impairments. Therefore, the implications of	
Clotting disorders: prothrombin elevation, thrombocytopenia	3	these disruptions must be evaluated and related to the individual and his pattern of	
Wernicke-Korsakoff syndrome	2	alcoholism. The following behavior patterns	
Alcoholic cerebellar degeneration	1	show psychological dependence on alcohol in alcoholism:	
Cerebral degeneration in absence of Alzheimer's disease or arteriosclerosis	2	1. Drinking despite strong medical contraindication known to patient	1
Central pontine myelinolysis ⎫ diagnosis ⎬ only Marchiafava-Bignami's ⎪ possible disease ⎭ postmortem	2 2	2. Drinking despite strong, identified, social contraindication (job loss for intoxication, marriage disruption because of drinking, arrest for intoxica-	
Peripheral neuropathy (see also beriberi)	2	tion, driving while intoxicated)	1
Toxic amblyopia	3	3. Patient's subjective complaint of loss	
Alcohol myopathy	2	of control of alcohol consumption	2
Alcoholic cardiomyopathy	2		
Beriberi	3		
Pellagra	3		

rectly diagnosed in the elderly individual may be very low (Gordon, et al., 1976).

The argument might be made that the elderly alcoholic is simply a person who has had a drinking problem for several years. Regardless of changes associated with aging, this individual would still manifest symptoms of chronic alcoholism as defined by criteria such as those developed by the National Council on Alcoholism (1972). This argument makes two assumptions concerning the elderly alcoholic. The first assumption is that the elderly alcoholic is the young alcoholic who has aged. The second assumption is that the clinician, e.g., psychologist, physician, psychiatric social worker, providing treatment can distinguish changes associated with aging from symptoms associated with alcohol abuse in the elderly (Gordon et al., 1976).

Is the elderly alcoholic a young alcoholic who has aged? A number of studies indicate that this may be only partially correct

Table 2

Major Criteria for the Diagnosis of Alcoholism

CRITERION	DIAGNOSTIC LEVEL	CRITERION	DIAGNOSTIC LEVEL
TRACK I. PHYSIOLOGICAL AND CLINICAL		Serum osmolality (reflects blood alcohol levels); every 22.4 increase over 200 mOsm/liter reflects 50 mg./ 100 ml. alcohol	2
A. Direct Effects (ascertained by examination)			
1. Early:			
Odor of alcohol on breath at time of medical appointment	2	3. Minor—Indirect	
2. Middle:		Results of alcohol ingestion:	
Alcoholic facies	2	Hypoglycemia	3
Vascular engorgement of face	2	Hypochloremic alkalosis	3
Toxic amblyopia	3	Low magnesium level	2
Increased incidence of infections	3	Lactic acid elevation	3
Cardiac arrhythmias	3	Transient uric acid elevation	3
Peripheral neuropathy (see also Major Criteria, Track I, B)	2	Potassium depletion	3
		Indications of liver abnormality:	
3. Late (see Major Criteria, Track I, B)		SGPT elevation	2
B. Indirect Effects		SGOT elevation	3
1. Early:		BSP elevation	2
Tachycardia	3	Bilirubin elevation	2
Flushed face	3	Urinary urobilinogen elevation	2
Nocturnal diaphoresis	3	Serum A/G ration reversal	2
2. Middle:		Blood and blood clotting:	
Ecchymoses on lower extremities, arms, or chest	3	Anemia: hypochromic, normocytic, macrocytic, hemolytic with stomatocytosis, low folic acid	3
Cigarette or other burns on hands or chest	3	Clotting disorders: prothrombin elevation, thrombocytopenia	3
Hyperreflexia, or if drinking heavily, hyporeflexia (permanent hyporeflexia may be a residuum of alcoholic polyneuritis)	3	ECG abnormalities	
		Cardiac arrhythmias: tachycardia; T waves dimpled, cloven, or spinous; atrial fibrillation; ventricular premature contractions; abnormal P waves	2
3. Late:			
Decreased tolerance	3	EEG abnormalities	
		Decreased or increased REM sleep, depending on phase	3
C. Laboratory Tests		Loss of delta sleep	3
1. Major—Direct		Other reported findings	3
Blood alcohol level at any time of more than 300 mg./100 ml.	1	Decreased immune response	3
Level of more than 100 mg./100 ml. in routine examination	1	Decreased response to Synacthen test	3
2. Major—Indirect		Chromosomal damage from alcoholism	3

(Rathbone-McCuan and Bland, 1975; Rosin and Glatt, 1971; Schuckit and Miller, 1976; Zimberg, 1974). Rosin and Glatt (1971) have suggested that elderly alcoholics can be differentiated, depending on the factors which precipitated their excessive drinking. The first type of elderly alcoholic (defined here as the chronic alcoholic) is the person who has had a long-standing problem with alcohol. Excessive drinking by these individuals is thought to be due to what Rosin and Glatt describe as "primary factors" such as the individual's personality and resulting behavioral patterns. Generally, the chronic alcoholics show personality characteristics similar to those frequently seen in younger alcoholics (egocentric, compulsive, self-indulgent). The second type of elderly alcoholic described by Rosin and Glatt is the individual who develops a drinking problem in response to "reactive factors" such as the loss of a spouse, retirement or impending retirement, and loneliness (defined here as the situational alcoholic). Excessive drinking in these individuals is thought to be precipitated by the physiological and psychological trauma sometimes associated

Table 2 (cont'd)

CRITERION	DIAGNOSTIC LEVEL	CRITERION	DIAGNOSTIC LEVEL
TRACK II. BEHAVIORAL, PSYCHOLOGICAL, AND ATTITUDINAL		B. Psychological and Attitudinal	
A. Behavioral		1. Direct effects	
1. Direct effects		Early:	
Early:		When talking freely, makes frequent reference to drinking alcohol, people being "bombed," "stoned," etc., or admits drinking more than peer group	2
Gulping drinks	3		
Surreptitious drinking	2		
Morning drinking (assess nature of peer group behavior)	2		
Middle:		Middle:	
Repeated conscious attempts at abstinence	2	Drinking to relieve anger, insomnia, fatigue, depression, social discomfort	2
Late:		Late:	
Blatant indiscriminate use of alcohol	1	Psychological symptoms consistent with permanent organic brain syndrome (see also Major Criteria, Track I, B)	2
Skid Row or equivalent social level	2		
2. Indirect effects			
Early:		2. Indirect effects	
Medical excuses from work for variety of reasons	?	Early:	
Shifting from one alcoholic beverage to another	2	Unexplained changes in family, social, and business relationships; complaints about wife, job, and friends	3
Preference for drinking companions, bars, and taverns	2	Spouse makes complaints about drinking behavior, reported by patient or spouse	2
Loss of interest in activities not directly associated with drinking	2	Major family disruptions: separation, divorce, threats of divorce	3
Late:		Job loss (due to increasing interpersonal difficulties), frequent job changes, financial difficulties	3
Chooses employment that facilitates drinking	3		
Frequent automobile accidents	3	Late:	
History of family members undergoing psychiatric treatment; school and behavioral problems in children	3	Overt expression of more regressive defense mechanisms: denial, projection, etc.	3
Frequent change of residence for poorly defined reasons	3	Resentment, jealousy, paranoid attitudes	3
Anxiety-relieving mechanisms, such as telephone calls inappropriate in time, distance, person, or motive (telephonitis)	2	Symptoms of depression: isolation, crying, suicidal preoccupation	3
Outbursts of rage and suicidal gestures while drinking	2	Feelings that he is "losing his mind"	2

with aging. It has been estimated that the situational alcoholic
compromises approximately 15% or 16% of the total number of alcoholics 55 years of age and older (Gordon et al., 1976). However,
because of the difficulty in identifying the elderly alcoholic
this percentage may be greatly underestimated.

If we assume that there are two major types of elderly alcoholics, how can they be differentially diagnosed? This is a particularly important question because of the differences in treatment for these two types of alcoholics (Rathbone-McCuan and Bland,
1975; Zimberg, 1974). Treatment for the situational alcoholic is
directed toward helping the person cope with changes in his or
her lifestyle. On the other hand, the treatment focus for the
chronic alcoholic involves more conventional treatment approaches,
e.g., aversion therapy, psychotropic drugs, halfway houses, Alcoholics Anonymous. The major distinction between the chronic alcoholic and the situational alcoholic is that the chronic alcoholic
shows evidence of psychopathological causes for excessive drinking, whereas the situational alcoholic's excessive drinking is
due in a large part to changes in psychological and/or physiological status (Rosin and Glatt, 1971). The chronic alcoholic is the
individual who shows a high incidence of psychopathology, particularly depressive and psychopathic traits. Also, it would be expected that the chronic alcoholic would have many of the symptoms
listed in Tables 1 and 2. For example, the chronic alcoholic may
have a history of employment problems as well as problems with
the police and marital instability, i.e., multiple marriages.
In comparison, the situational alcoholic may show few, if any,
of the symptoms listed in Tables 1 and 2, other than excessive
drinking. For these individuals, abuse of alcohol may have begun
when many of the Track II criteria are no longer representative of
normal functioning. Also, some of the physical consequences of
prolonged drinking may not have developed in this group. Other
factors which may differentiate the chronic alcoholic from the
situational alcoholic include: frequency and amount of alcohol
used, type of alcoholic beverage consumed, and the effects of
alcohol on the individual's performance (Rathbone-McCuan and
Bland, 1975). For the most part the preceding factors would
probably be fairly stable for the chronic alcoholic but would show
a significant change for the situational alcoholic.

One point worth noting is that "frequency" and "amount" reports should be interpreted cautiously in attempts to diagnose
alcoholism in the elderly individual. Elderly individuals show a
decreased tolerance to alcohol (Rosin and Glatt, 1971). Moreover, elderly individuals exhibit a higher blood level of alcohol
than younger persons when administered equal doses of the drug
(Vestal et al., 1977). The clinical significance of these findings is that it takes less alcohol to produce an effect in an

elderly person when compared to a younger individual. Even though
the elderly individual may consume a smaller amount of alcohol
than is generally associated with alcohol abuse, the amount reported may well be excessive because of the augmented effect of
alcohol for the elderly.

Alcoholics with Organic Brain Syndrome

The actual incidence of organic brain syndrome (OBS) in the
elderly population is not known. Studies that have examined various medical and psychiatric populations in institutionalized settings have reported the occurrence of this disorder ranging from
25% (Schuckit and Miller, 1976) to as high as 61% (Gaitz and Baer,
1971). These percentages probably do not reflect the prevelance
of elderly alcoholics with OBS in the community. It would be
expected that there would be more OBS alcoholics in institutions
than there are in the community (Gaitz and Baer, 1971).

Prolonged and excessive consumption of alcohol can result in
pathological changes in the central nervous system of some individuals. Chronic alcoholism can also produce pathological changes
in the peripheral nervous system, e.g., peripheral neuropathy;
however, this discussion is confined to alcoholics with the primary diagnosis of alcohol-induced OBS. Several central neurological diseases have been attributed to excessive alcohol consumption,
e.g., cerebellar cortical degeneration, central pontine myelinoly-
sis, Marchiafava-Bignami disease, and cerebral degeneration, with
the most common being Wernicke-Korsakoff syndrome (Dreyfus, 1974).
While various factors have been proposed to explain the neurological damage associated with chronic alcoholism, it has been
suggested (Neville, Eagles, Samson, and Olson, 1968; Victor and
Adams, 1961) that nutritional deficiency may be one of the primary causes of many of these conditions. Wernicke-Korsakoff
syndrome clearly appears to be due to a severe depletion of
Vitamin B1, or Thiamine (Victor and Adams, 1961; Victor, Adams,
and Collins, 1971). The elderly in general are very susceptible
to nutritional deficiencies (Cain, 1977) and the elderly alcoholic is even more susceptible.

Brain damage also exhibits an increase with advancing age in
the general population (Verwoerdt, 1976). It has been estimated
that at least one out of every four elderly alcoholics may have
some type of brain damage (Schuckit and Miller, 1976; Simon et al.,
1968). Given that chronic alcoholism can result in brain damage
and that brain damage is more prevalent in the elderly population, how can the elderly alcoholic with OBS be differentiated
from the elderly nonalcoholic with OBS? Of the few studies that
have been conducted (Gaitz and Baer, 1971; Schuckit and Miller,
1976); Simon et al., 1968) the general finding is that elderly OBS

alcoholics are less impaired on several dimensions when compared
to elderly OBS nonalcoholics. The OBS alcoholics show fewer
psychiatric symptoms than the OBS nonalcoholics, e.g., hallucinations, delusions, social withdrawal. The OBS alcoholics generally
perform better on a variety of measures such as the Wechsler
Memory Scale and some tests of the Halstead-Reitan Neuropsychological Test Battery. Demographically, the OBS alcoholics are
younger and are likely to report multiple marriages. Even though
the OBS alcoholics are generally younger than the OBS nonalcoholics, the mortality rate appears to be approximately the same
for both groups (Gaitz and Baer, 1971). This would suggest that
the OBS alcoholics may have more serious health problems than the
OBS nonalcoholics.

Although there are differences between these two OBS groups,
differential diagnosis in some instances may be difficult because
chronic alcoholism also can be complicated by senile or arteriosclerotic brain damage (Gaitz and Baer, 1971). An attempt should
be made to obtain an accurate drinking history from relatives or
close friends if it is suspected that the OBS may be alcohol-
induced. It is important to distinguish between these two OBS
groups in order that proper treatment can be provided. For example, Wernicke-Korsakoff syndrome is thought to be due to a severe
vitamin deficiency. Some individuals respond very favorably to a
treatment regime utilizing thiamine (Victor and Adams, 1961).

The elderly OBS alcoholic also can be compared with the elderly alcoholic without OBS. However, as with the differentiation between alcoholic and nonalcoholic OBS groups, our knowledge
is preliminary and tentative. It has been observed that the OBS
alcoholic is younger (Schuckit and Miller, 1976), although some studies
have reported no differences in age (Gaitz and Baer, 1971), and
others have reported that the OBS alcoholic is actually older than
the non-OBS alcoholic (Simon et al., 1968). As would be expected,
the OBS alcoholics are more cognitively impaired and exhibit more
psychiatric symptoms than alcoholics without signs of OBS (Gaitz
and Baer, 1971; Schuckit and Miller, 1976; Simon et al., 1968).
One of the more outstanding differences between the two groups
of alcoholics is health status. Non-OBS alcoholics have fewer
serious health problems and have a lower mortality rate. OBS
alcoholics also have a tendency to live alone and have fewer employment and non-traffic police problems than do the alcoholics
without OBS (Schuckit and Miller, 1976).

CONCLUSIONS AND SUMMARY

Alcoholism is a serious problem in the elderly population.
The situation is complicated by the special physical and psycho-

logical problems which are sometimes associated with aging. Moreover, alcoholism in the elderly person has not been regarded as a
problem. In some instances the elderly alcoholic has been considered a very poor candidate for treatment. In the national survey of agencies working with alcoholics, Gordon and her colleagues
found that 32% of the agencies responding did not believe that
the elderly alcoholic required special consideration. Some respondents viewed the elderly alcoholic "as a virtually hopeless
and incurable group and consequently, not one for which special
efforts would be beneficial" (Gordon et al., 1976, p. 39). On
the other hand, Schuckit and Miller have pointed out "that any
facility that has good medical care and the opportunity to offer
social work help, along with occupational and/or recreational therapy, should be able to deal effectively with the geriatric alcoholic" (Schuckit and Miller, 1976, p. 559). However, this statement assumes that the clinician is aware that alcoholism is a
problem in the elderly population. Generally this does not appear
to be the case (Gordon et al., 1976). Of the agencies in the
national survey that realized that alcoholism can be a problem in
the elderly individual, only 10% had any special programs for
these individuals. It would appear that there is an immediate need
to increase the clinician's and the public's awareness of the fact
that alcohol abuse is a real problem in the elderly population.

One way of increasing the awareness that alcohol abuse is a
problem for the elderly is through the education and training of
physicians. It has been estimated that one-third of non-institutionalized individuals over 65 come in contact with a physician
during a four-week period (Rathbone-McCuan and Bland, 1975). It
would seem that the physician is in a key position to identify
alcohol abuse in the elderly individual. However, "physicians
often are unable or unwilling to recognize alcohol abuse in their
patients" (Zimberg, 1974, p. 223). Also, the elderly alcoholic
may be viewed as "incurable" because of aging and alcoholism and
consequently little time may be spent in trying to help these
individuals (Rathbone-McCuan and Bland, 1975).

The primary emphasis of this chapter has been on the identification of the elderly alcoholic. Diagnostically, it would
seem that the elderly alcoholic can be differentiated on the basis
of whether primary or reactive factors precipitated the alcohol
abuse. An awareness of this distinction is important in regard to
treatment as well as in identifying elderly situational alcoholics
who may be "caught up" in the variety of treatment services for
the elderly for problems that are actually caused by alcohol
abuse. Interestingly, not only does the diagnostic classification
apply to elderly alcoholics but a similar kind of classification
has been proposed for alcoholics in general (Knight, 1937; Pattison, 1974; Reilly and Sugerman, 1967). Knight (1937) has classified alcoholics as being either "essential alcoholics," with a

deficiency in basic coping skills, or "reactive alcoholics" with
normally sufficient coping skills which are inadequate when confronted with overwhelming stress. It is possible that the two
types of elderly alcoholics correspond with those described by
Knight. If this is true, knowledge of an individual's type of
drinking patterns over the life-span might serve as a clue in
predicting and preventing excessive drinking when these individuals are older. Granted, this hypothesis is speculative; however,
it is safe to say that to a great extent our knowledge concerning
alcohol abuse in the aged individual is still at a speculative
and tentative level.

REFERENCES

Bailey, M. B., Haberman, P. W., and Alksne, H. The epidemiology
    of alcoholism in an urban residential area. Quarterly Jour-
    nal of Studies on Alcohol, 1965, 26, 19-40.
Barry, H. Psychological factors in alcoholism. In B. Kissin and
    H. Begleiter (Eds.), The Biology of Alcoholism (Vol. 3).
    New York: Plenum Press, 1974.
Burch, G. E. and Giles, T. D. Alcoholic cardiomyopathy. In B.
    Kissin and H. Begleiter (Eds.), The Biology of Alcoholism
    (Vol. 3). New York: Plenum Press, 1974.
Cahalan, D. and Cisin, I. H. Drinking behavior and drinking prob-
    lems in the United States. In B. Kissin and H. Begleiter
    (Eds.), The Biology of Alcoholism (Vol. 4). New York:
    Plenum Press, 1976.
Cain, L. D. Evaluative research and nutrition programs for the
    elderly. In Evaluative Research and Social Programs for the
    Elderly. DHEW Publication No. 20120. Washington, D.C.:
    U.S. Government Printing Office, 1977.
Cermak, L. S. and Ryback, R. S. Recovery of verbal short-term
    memory in alcoholics. Journal of Studies on Alcohol, 1976,
    37, 46-52.
Criteria Committee, National Council on Alcoholism. Criteria for
    the diagnosis of alcoholism. American Journal of Psychiatry,
    1972, 129, 127-135.
de Lint, J. and Schmidt, W. Alcoholism and mortality. In B.
    Kissin and H. Begleiter (Eds.), The Biology of Alcoholism
    (Vol. 4). New York: Plenum Press, 1976.
Drew, L. R. Alcoholism as a self-limiting disease. Quarterly
    Journal of Studies on Alcohol, 1968, 29, 956-967.
Dreyfus, P. M. Diseases of the nervous system in chronic alco-
    holics. In B. Kissin and H. Begleiter (Eds.), The Biology
    of Alcoholism (Vol. 3). New York: Plenum Press, 1974.
Ernst, A. J., Dempster, J. P., Yee, R., St. Dennis, C., and Nakano,
    L. Alcohol toxicity, blood alcohol concentration and body

water in young and adult rats. *Journal of Studies on Alcohol*, 1976, 37, 347-356.

Feinman, L. and Lieber, C. S. Liver disease in alcoholism. In B. Kissin and H. Begleiter (Eds.), *The Biology of Alcoholism* (Vol. 3). New York: Plenum Press, 1974.

Fisher, J. C., Mason, R. L., and Fisher, J. V. A diagnostic formula for alcoholism. *Journal of Studies on Alcohol*, 1976, 37, 1247-1255.

Friedel, R. O. and Raskind, M. A. Psychopharmacology of aging. In M. F. Elias, B. E. Eleftheriou, and P. K. Elias (Eds.), *Special review of experimental aging research: Progress in Biology*. Bar Harbor, Maine: EAR Inc., 1976.

Gaitz, C. M., and Baer, P. E. Characteristics of elderly patients with alcoholism. *Archives of General Psychiatry*, 1971, 24, 372-378.

Goldberg, P. B. and Roberts, J. Influence of age on the pharmacology and physiology of the cardiovascular system. In M. F. Elias, B. E. Eleftheriou, and P. K. Elias (Eds.), *Special Review of Experimental Aging Research: Progress in Biology*. Bar Harbor, Maine: EAR Inc., 1976.

Goodrick, C. L. Behavioral characteristics of young and senescent inbred female mice of the C57BL/6J strain. *Journal of Gerontology*, 1967, 22, 459-464.

Goodrick, C. L. Behavioral differences in young and aged mice: Strain differences for activity measures, operant learning, sensory discrimination, and alcohol preference. *Experimental Aging Research*, 1975, 1, 191-207.

Gordon, J. J., Kirchoff, K. L., and Philipps, B. K. *Alcoholism and the Elderly*. Iowa City: Elderly Program Development Center, Inc., 1976.

Hillman, R. W. Alcoholism and malnutrition. In B. Kissin and H. Begleiter (Eds.), *The Biology of Alcoholism* (Vol. 3). New York: Plenum Press, 1974.

Jacobson, G. R. *The Alcoholisms: Detection, Diagnosis and Assessment*. New York: Human Sciences Press, 1976.

Jellinek, E. M. *The Disease Concept of Alcoholism*. New Haven: Hillhouse Press, 1960.

Kissin, B. The pharmacodynamics and natural history of alcoholism. In B. Kissin and H. Begleiter (Eds.), *The Biology of Alcoholism* (Vol. 3). New York: Plenum Press, 1974.

Klisz, D. K. and Parsons, O. A. Hypothesis testing in younger and older alcoholics. 1977, in press.

Knight, R. P. The dynamics and treatment of chronic alcohol addiction. *Bulletin of the Menninger Clinic*, 1937, 1, 233-250.

Lorber, S. H., Dinoso, V. P., and Chey, W. Y. Diseases of the gastrointestinal tract. In B. Kissin and H. Begleiter (Eds.), *The Biology of Alcoholism* (Vol. 3). New York: Plenum Press, 1974.

Lyons, H. A. and Saltzman, A.   Diseases of the respiratory tract
    in alcoholics.   In B. Kissin and H. Begleiter (Eds.), The
    Biology of Alcoholism (Vol. 3).   New York: Plenum Press,
    1974.

Neville, J. N., Eagles, J. A., Samson, G., and Olson, R. E.   Nutri-
    tional status of alcoholics.   American Journal of Clinical
    Nutrition, 1968, 21, 1329-1340.

Omenn, G. S.   Pharmacogenetics and aging.   In M. F. Elias, B. E.
    Eleftheriou, and P. K. Elias (Eds.), Special Review of
    Experimental Aging Research: Progress in Biology.   Bar
    Harbor, Maine: EAR Inc., 1976.

Pattison, R. D., Abrahams, R., and Baker, F.   Preventing self-
    destructive behavior.   Geriatrics, 1974, 29, 115-121.

Pattison, E. M.   Rehabilitation of the chronic alcoholic.   In
    B. Kissin and H. Begleiter (Eds.), The Biology of Alcoholism
    (Vol. 3).   New York: Plenum Press, 1974.

Rathbone-McCuan, E., and Bland, J.   A treatment typology for the
    elderly alcohol abuser.   Journal of the American Geriatrics
    Society, 1975, 23, 553-557.

Reilly, D. H. and Sugerman, A. A.   Conceptual complexity and
    psychological differences in alcoholics.   Journal of Ner-
    vous and Mental Diseases, 1967, 144, 14-17.

Rosin, A. J., and Glatt, M. M.   Alcohol excess in the elderly.
    Quarterly Journal of Studies on Alcohol, 1971, 32, 53-59.

Schuckit, M. A., and Miller, P. L.   Alcoholism in elderly men:
    A survey of a general medical ward.   Annals of the New York
    Academy of Sciences, 1976, 273, 558-571.

Shuster, L.   Age-related changes in pharmacodynamics of drugs that
    act on the central nervous system.   In M. F. Elias, B. E.
    Eleftheriou, and P. K. Elias (Eds.), Special Review of
    Experimental Aging Research: Progress in Biology.   Bar
    Harbor, Maine: EAR Inc., 1976.

Siassi, I. G., Crocetti, H., and Spiro, R.   Drinking patterns and
    alcoholism in a bluecollar population.   Quarterly Journal
    of Studies on Alcohol, 1973, 34, 917-926.

Simon, A., Epstein, L. J., and Reynolds, L.   Alcoholism in the
    geriatric mentally ill.   Geriatrics, 1968, 23, 125-131.

Sun, A. Y., Ordy, J. M., and Samorajski, T.   Effects of alcohol
    on aging in the nervous system.   In J. M. Ordy and K. R.
    Brizzee (Eds.), Neurobiology of aging: An interdisciplinary
    life-span approach.   New York: Plenum Press, 1975.

Timiras, P. S.   Developmental Physiology and Aging.   New York:
    Macmillan, 1972.

Verwoerdt, A.   Clinical Geropsychiatry.   Baltimore: The Williams
    & Williams Co., 1976.

Vestal, R. E., McGuire, E. A., Tobin, J. D., Andres, R., Norris,
    A. H., and Mezey, E.   Aging and ethanol metabolism.   Clini-
    cal Pharmacology and Therapeutics, 1977, 21, 343-354.

Victor, M. and Adams, R. D.   On the etiology of the alcoholic
    neurological diseases: With special referral to the role

of nutrition. <u>American Journal of Clinical Nutrition</u>, 1961, <u>9</u>, 379-397.

Victor, M., Adams, R. D., and Collins, G. H. The Wernicke-Korsa-koff Syndrome: A clinical and pathological study of 245 patients, 82 with postmortem examinations. Philadelphia: F. A. Davis, 1971.

Waller, J. A. Alcohol and unintentional injury. In B. Kissin and H. Begleiter (Eds.), <u>The Biology of Alcoholism</u> (Vol. 4). New York: Plenum Press, 1976.

Wiberg, G. S., Trenholm, H. L., and Coldwell, B. B. Increased ethanol toxicity in old rats: Changes in LD50, <u>in vivo</u> and <u>in vitro</u> metabolism and liver alcohol dehydrogenase activ-ity. <u>Toxicology and Applied Pharmacology</u>, 1970, <u>16</u>, 718-727.

Wood, W. G. Age-associated differences in response to alcohol in rats and mice: A biochemical and behavioral review. <u>Experi-mental Aging Research</u>, 1976, <u>2</u>, 543-562. (a)

Wood, W. G. Ethanol preference in the C57BL/6 and BALB/c mice at three ages and eight ethanol concentrations. <u>Experimental Aging Research</u>, 1976, <u>2</u>, 425-434. (b)

Wood, W. G., Elias, M. F., Schultz, N., Dineen, J., and Pentz, C. A. Differences in performance between younger and older alcoholics on personality, neuropsychological, and informa-tion processing tests. Unpublished manuscript, Syracuse Uni-versity, 1977.

Zimberg, S. The elderly alcoholic. <u>The Gerontologist</u>, 1974, <u>14</u>, 221-224.

Section 2
Personality Assessment

INTRODUCTION TO SECTION 2: PERSONALITY ASSESSMENT

Ilene C. Siegler

Duke University Medical Center

The chapters is this section deal with the assessment of personality. Objective measures of personality are discussed by Costa and McCrae and projective measures by Kahana. Gentry's chapter on issues in psychosomatic assessment discusses the nature of the relationships between psychological factors and physical illness. Each chapter contains reviews of the relevant literature, describes the clinical relevance and practical applications of the techniques reviewed, and provides suggestions for future research. However, the chapters are very different in terms of the degree of sophistication of assessment procedures described due to the widely varying degrees of existing literature in the three topic areas.

Costa and McCrae's chapter on objective personality assessment emphasizes that the study of personality in older persons rightfully belongs in the context of adult development. New data are presented which suggest that personality is relatively stable during the adult years; observed changes need to be separated into those due to cultural changes (cohort effects) and those due to maturation (age effects). Perhaps cohort norms rather than age norms make better sense in the assessment of older adults. However, an age group such as the elderly often spans two or more generations. Thus, seven to ten year cohorts useful for research purposes may have little translation value in practical terms. We do not yet know what the optimal range should be. Further, many investigators see cohort as a variable similar to age in that "cohort" merely points the way to the necessary specification of the "real" variables that are causing an effect.

Openness to experience in later life presents an important area for future research. While there has been much research concerned with cautiousness and rigidity, openness to experience provides the other side of what may be a bipolar phenomenon. The meaning and measurement of various indices of quality of life such as morale, life satisfaction, and happiness may also be important foci.

The data on midlife crisis presented by Costa and McCrae represent an important first step in providing an empirical test of current crisis-oriented adult development theory. The authors are careful to point out that their data pertain to men only. It should be recognized that sex differences as well as age and/or cohort by sex interactions may be important in personality assessment. It is clear that those who are now older adults have undergone socialization experiences that are clearly sex typed. Future studies that seek to explicate age differences between generations need to be aware of potential age by sex interactions. However, sex differences in old age may be less distinct in the future due to such factors as increases in life expectancy, the increasing employment rate of women, and a greater emphasis on smaller families.

Kahana's chapter on projective assessment techniques and their use with older persons is written in a practical vein. He argues that projective techniques are particularly appropriate for older adults in that factors such as social class, educational level, sensory deficits, and missing data can be evaluated during the assessment procedure and thus become useful data. Kahana points out an important gap in the research literature on the use of projective tests with older adults: the majority of the validation studies have examined the influence of cognitive and sensory factors on test performance, rather than the usefulness of projective techniques in conjunction with other measures of personality and adjustment. As do Costa and McCrae, Kahana emphasizes the positive aspects of personality and ego strength in later life and calls for a more humanistic psychology of old age.

Gentry's chapter on issues in psychosomatic assessment serves to orient both researchers and clinicians to the appropriate questions to be asked about the relationships between physical illness and symptomatology as well as psychological reactions and symptoms. His chapter points out issues in both the diagnosis and treatment of conditions common to older adults where physical symptoms are likely to be present. This is an area where psychologists have done little but have a major contribution to make.

OBJECTIVE PERSONALITY ASSESSMENT[1]

Paul T. Costa, Jr., Ph.D.

Normative Aging Study, V. A. Outpatient Clinic
and
University of Massachusetts at Boston

Robert R. McCrae, Ph.D.

University of Massachusetts at Boston

This chapter aims to provide a broad, empirically-based sketch of normal personality and its changes throughout adulthood. While it is not intended as a cookbook treatment of the application of objective tests to geriatric populations, it does attempt to make specific links between normal personality and clinical gerontology. One example of this is provided by a consideration of the male mid-life crisis, that putatively universal, normative, developmental phase which at the same time can be considered a pre-geriatric clinical condition. Similarly, our theoretical discussion of openness to experience—a personality dimension relevant to the call for more humanistic approaches to aging—is supplemented by presentation of an instrument which may prove a useable operationalization of it. Objective personality assessment has a long and controversial past, but evidence and arguments presented here suggest that its careful use can make a substantial contribution to the field.

Within the field of psychological measurement, the term "objective" can refer to performance measures like the Embedded Figures Test, to self-report scales, or to almost any non-projective form of assessment. Classifications of types of tests, together with discussions of their particular strengths and weaknesses, are

[1]Research supported in part by the Council for Tobacco Research—U.S.A., Inc., Grant #1085R1 (University of Massachusetts at Boston) and the Medical Research Service of the Veterans Administration (Normative Aging Study).

offered by Fiske (1971), Cattell (1946), and others. Standard arguments for objective measures point out that objective scoring systems eliminate both the necessity for clinical expertise in interpretation and the possibility of clinical bias. Quantification,
which is a hallmark of objective tests, also permits rigorous
assessment of reliability and validity, and makes possible sophisticated statistical treatment of group comparisons.

Performance measures of personality, while appealing as the
apparent ultimate in objectivity, will not be reviewed in this
chapter since they are at present quite limited in their utility
(see Hundleby, 1973). Chown's (1961) use of behavioral measures
of rigidity illustrates how performance measures can supplement
self-report instruments, but the latter will continue, for the
near future, to be the chief source of objective data on adult
personality. The remainder of this chapter will therefore deal
only with self-report measures. It is not intended to be an exhaustive review of the literature on the objective measurement of
adult personality, since other recent reviews have accomplished
this (Schein, 1968; Chown, 1968; Schaie and Marquette, 1972;
Neugarten, 1977); nor will it focus on strictly clinical instruments like the MMPI. Instead, new data on a large sample of normal adults will be used to view clinical gerontology in the larger
context of adult personality and development.

Specifically, the three sections of the chapter will address
three separate issues. The first reports longitudinal data on
stability and change in normal personality. In the absence of
good empirical norms, evidence on the nature and magnitude of personality change in normal individuals can provide the clinician
(whose direct experience may be limited to pathological cases)
with an intuitive baseline against which to judge the condition of
his or her clients. Moreover, it can provide a corrective to
pathology-based theories of aging. The second section deals with
a broad domain of personality, openness to experience, which has
important implications for questions of personality development,
well-being, and the quality of life. Data on the development and
validation of an instrument to measure several facets of the domain are presented, and age trends in the level of openness are
discussed. Finally, the third section presents empirical data on
a "new" condition of interest to clinicians, the "mid-life crisis."

On a more general level, this chapter can be taken as a kind
of defense of self-report methodology. Confidence in this form of
assessment has been eroded in the last few decades by concern for
response sets like acquiescence and social desirability (Crowne
and Marlowe, 1964), and by attacks on the utility of trait measures
in predicting behavior (Mischel, 1968). Additionally, gerontologists properly have been hesitant to assume that personality

measures, almost uniformly developed on college students and young adults, can be meaningfully applied to older subjects. The evidence presented here supports the more recent arguments (Hogan, DeSoto, and Solano, 1977) that self-report methods still have a major contribution to make to the understanding of adult personality.

Readers who are interested in a comprehensive review of the clinical use of objective tests in the elderly are referred to Lawton, Whelihan, and Belsky (in press). Lawton and his colleagues give a detailed review of four objective tests: the MMPI, 16PF, Edwards Personal Preference Schedule, and the Maudsley Personality Inventory. They recommend, although with reservations, the Mini-Mult as a useable reduced form of the MMPI for clinical assessment of the elderly, and the Maudsley and 16PF as appropriate measures for personality description. Lawton discusses the difficulties which cognitive and sensory deficits present for many older people in taking objective tests, and encourages a more personalized approach on the part of the tester to overcome motivational problems. He summarizes information on the availability of norms and on validity evidence for major personality tests and concludes that "the state of the art is not encouraging." This pessimistic view is derived not from evidence of the failure of objective instruments in elderly populations, but from the general lack of relevant research. Some of the literature on objective testing in the elderly is presented briefly here, but the present chapter focuses primarily on presenting some of that research evidence which Lawton rightly demands.

THE STABILITY OF OBJECTIVELY ASSESSED PERSONALITY

A large body of studies on age differences in the level of personality variables has by now been accumulated. Neugarten's (1977) recent review cites findings of cross-sectional differences in such diverse areas as rigidity, locus of control, and dependency, but she concludes that stability in personality is the rule. The only finding of change which is consistently replicated is an increase of introversion in the second half of life.

Some have looked at cross-sectional differences in MMPI scales. Calden and Hokanson (1959) reported significantly higher scores for older subjects on Hypochondriasis, Depression, and Social Introversion, and advocated the establishment of age-norms for interpreting MMPI scores. Their results were replicated on normal and clinical populations by Swenson (1961) and by Aaronson (1958). Other studies (Brozek, 1955; Hardyck, 1964) report that older subjects have lower Hypomania scores and higher Hypochondriasis and Hysteria scores, while lower F scores were found by

Gynther and Shimkumas (1965) with psychiatric patients. As Aaronson concludes, older subjects seem to be less troubled by impulse control and more concerned with mental and physical health.

Since all these studies are cross-sectional, they do not answer the question of whether differences are truly developmental or simply cohort effects. This question is of practical as well as theoretical importance. For example, if effects are not developmental, then suggestions for providing age norms would be inappropriate; instead, cohort norms might be more meaningful. If level differences in a variable are developmental in origin, then the appropriate comparison group for evaluation of an individual's score is provided by age norms, i.e., scores of individuals of similar chronological age. However, if effects are due to generational differences, then the appropriate comparison group will be provided by individuals of similar birth year. The age and cohort comparison groups will, of course, be the same when contemporaneously-gathered norms are used. But when, as is typically the case, clinicians must use norms collected some years ago, the two comparison groups may be quite different. If, for example, age differences in level of MMPI scales scores are due solely to cohort or generational effects, then the data which Calden and Hokanson gathered on 50-year-olds in 1959 should be applied to today's 70-year-olds, not today's 50-year-olds. The issue is further complicated by the fact that the meaning of test items may change with time and the longevity of test norms is largely unknown.

In an attempt to answer such questions, elaborate designs for separating the effects of cohort, aging, and cultural change have been proposed (Schaie, 1965), but the difficulty of implementing them has meant that they have rarely been used on a large scale. Perhaps the most comprehensive and ambitious project to date examined scales from the Guilford-Zimmerman Temperament Survey (GZTS; Guilford, Zimmerman, and Guilford, 1976) on a sample of 930 men over a 14-year period. Douglas and Arenberg (1977) report maturational declines in General Activity (after age 50) and Masculinity, and generational differences in Restraint and Ascendence. Since both activity and ascendence are components of extraversion, the often-reported decline in extraversion may thus be a compound of both maturational and generational effects.

Our own data on personality stability is based on responses to the Sixteen Personality Factor Questionnaire (16PF; Cattell, Eber, and Tatsuoka, 1970) from participants in the Normative Aging Study (Bell, Rose, and Damon, 1972). This sample is composed of over 2000 male volunteers, mostly veterans, who were screened at entry for physical health and geographic stability. All but the lowest socio-economic levels are well-represented, and, in comparison to the general population, higher educational levels are found in the oldest men: in consequence, many commonly reported cohort effects which are linked to change in educational levels over the past half century are <u>not</u> found in the Normative Aging Study.

Since the primary focus in the Normative Aging Study was originally on physical health and aging, a regular cycle of psychological measurement was not planned, and time-sequential and cross-sequential designs were not employed. Instead, a one-time battery was administered to about 1100 subjects between 1965 and 1967 which included the 16PF (combined A and B forms, 1962 edition), the Strong Vocational Interest Blank (SVIB), the Allport-Vernon-Lindsay Scale of Values (AVL), and the General Aptitude Test Battery (GATB). Factor analyses (Costa, Fozard, and McCrae, 1977; Costa, Fozard, McCrae, and Bosse´, 1976) of SVIB and GATB scales were used to reduce the number of variables to more manageable proportions.

Between 1975 and 1977 a number of new personality measures were mailed to subjects, including the Crowne-Marlowe Social Desirability Scale (Crowne and Marlowe, 1964), the Traditional Family Ideology scale (Levinson and Huffman, 1955), the Eysenck Personality Inventory (EPI; Eysenck and Eysenck, 1968), the Experience Inventory (Coan, 1972) and scales to measure positive affect and the mid-life crisis. Data from all these measures will be mentioned in later sections of this chapter.

Of immediate interest are the longitudinal data provided by a readministration of the 16PF. In 1975, a subsample completed the 1962 edition of the A form of the 16PF, a revised edition in which about one-third of the original items had been replaced or rewritten. Direct comparison of the two sets of scales was impossible since it could not be determined whether results were due to longitudinal changes or to the presence of so many new and different items in the 1975 administration.

For thirteen of the scales this difficulty was overcome by using scores based on only those items which were found in both 1962 and 1967 editions. The remaining three scales--I (Tendermindedness), M (Imaginativeness), and Q1 (Liberal Thinking)--were of particular interest, and so the full 1962 A form of these three scales were readministered in 1977. Table 1 presents the correlations of the resulting scales over a ten-year interval, together with the number of items making up the scale. In general, these coefficients show considerable evidence of stability of individual differences in the aspects of personality measured by the 16PF. When longer, and thus more reliable, scales are used, measuring the broader dimensions of general anxiety and extraversion, stability coefficients reach the .80 mark (Costa and McCrae, in press).

Table 2 presents results of repeated measures analysis of all

16 scales, with three age cohorts as a second classifying variable. Cohort or cross-sectional age difference--but not longitudinal-- effects are seen for G (Conscientiousness), I (Tendermindedness), and Q1 (Liberal Thinking), showing an increase in all three across age groups. Effects for I and Q1 may reach significance in part because of the larger sample size for these variables. Longitudinal changes--but not cohort differences--are seen for B (Brightness) and Q2 (Independence), both showing an increase over time. There were no significant time-by-cohort interactions. The present design cannot separate cultural change from maturational in these latter cases, and, indeed, an artifact of testing may account for the apparent rise in intelligence. The scale was originally administered in groups with implicit time pressure. The second

Table 1

Ten-Year Stability of 16PF Scales

1962 "A" Scales[a]	1967 "A" Shared Items (N = 139)	1967 "A" Full Scales (N = 424)
A - Outgoing (5)	50	60
B - Bright (9)	24	45
C - Stable (10)	55	56
E - Assertive (11)	60	63
F - Happy-go-lucky (13)	64	71
G - Conscientious (7)	46	48
H - Adventurous (9)	61	74
I - Tender-minded (10)	63[b]	54
L - Suspicious (5)	39	40
M - Imaginative (13)	44[b]	16
N - Shrewd (3)	49	25
O - Guilt-prone (5)	51	55
Q1 - Liberal thinking (10)	46[b]	26
Q2 - Independent (6)	53	56
Q3 - Controlled (4)	32	47
Q4 - Tense (8)	64	67

    [a]Number of items shared by 1962 and 1967 editions given in parentheses.
    [b]Full 1962 "A" scales readministered 1977, N = 403. Decimal points omitted.

Table 2

Repeated Measures Means for 16PF Scales

Scale	Age Group			F for Age	F for Time
	25 - 40 (N = 46)	41 - 46 (N = 51)	47+ (N = 42)		
A — Outgoing	5.94	5.28	5.78	1.48	2.64
B — Bright	6.07	5.76	5.74	12.80	13.78***
C — Stable	7.04	5.91	6.50	1.93	0.59
E — Assertive	5.15	5.25	6.26	1.37	0.36
F — Happy-go-lucky	2.32	1.54	3.00	1.57	2.73
G — Conscientious	0.47	1.06	1.92	5.49**	2.59
H — Adventurous	1.17	0.52	0.49	0.54	0.21
I^a — Tender-minded	7.29	7.99	8.45	5.18**	1.27
L — Suspicious	2.85	2.69	2.55	0.30	2.94
M^a — Imaginative	11.98	12.03	12.53	1.92	2.48
N — Shrewd	1.97	2.42	2.63	2.96	0.06
O — Guilt-prone	1.19	1.76	1.31	1.28	0.05
Q1$_a$ — Liberal thinking	9.37	9.81	10.08	3.35*	0.96
Q2 — Independent	3.40	3.45	2.92	0.88	19.33***
Q3 — Controlled	1.08	0.59	1.07	1.83	0.04
Q4 — Tense	4.69	3.91	5.01	1.38	1.26

^{a}Based on full scale retest in 1977, Ns = 134, 101, and 169 for three age groups.

*p = .05; **p = .01; ***p = .001

measure was mailed to subjects and they completed it at leisure.
This additional time may be responsible for significant increases
in performance.

These results only partially confirm other reports of change.
The maturational decline of masculinity reported by Douglas and
Arenberg and paralleled by projective findings of Gutmann (1964)
and others is mirrored only in the cross-sectional differences
in I (Tendermindedness) in the present data; although an increase
in introversion is seen in Q2 (Independence), it is not seen in
any of the other 16PF scales (A, E, F, and H) measuring introversion-extraversion.

This lack of evidence for substantial maturational change
can be attributed at least in part to an inherent shortcoming in
longitudinal research. Cross-sectional designs can compare individuals across the full range of the adult life span; longitudinal studies typically cover only a few years and changes within
this interval may be simply too small to detect.

The major conclusion to be drawn about objectively measured
personality, at least in men, is that it shows remarkable stability of level throughout adulthood. There appear to be declines
in a few variables--extraversion and masculinity, for example--
but these changes are very small in comparison to the range of
variation among individuals in any cohort. The clinician may
expect, for example, some small increase in social withdrawal
with age, but a pronounced change is not a feature of normal aging
and may be a sign of psychopathology or a failure of the social
support systems.

The extreme aged, however, typified by multiple deficits in
cognitive, physical, and social functioning, have infrequently
been assessed in objective personality research because of the
special problems this group presents. Valid administration of
the tests may be hindered by poor hearing or vision and the physical endurance of these individuals severely limits testing sessions. However, a more fundamental problem which must be considered is the nature of personality itself among those characterized
by chronic brain syndrome or senile dementia. In such persons,
the appropriateness of traditional trait and state conceptions of
personality may not be tenable and "personality assessment" and
"personality change" may be meaningless terms. Similar difficulties are encountered in assessing the comatose or catatonic.

THE DOMAIN OF OPENNESS TO EXPERIENCE

In an earlier paper (Costa and McCrae, 1976) a cluster analysis

of the 16PF was reported which identified a third major dimension
of personality to supplement the long-established dimensions of
anxiety-adjustment and introversion-extraversion. In order to confirm our interpretation of this third cluster as representing a
dimension of experiential openness, it was necessary to measure
specific facets of openness with an accuracy not afforded by the
16PF scales. Fortunately, considerable work in this direction
had already been done by Coan and his colleagues in the development of an Experience Inventory (EI; Coan, 1972).

Conceptually, Coan's work had begun in the psychoanalytic
tradition. An attempt to measure objectively the capacity for
regression in service of the ego (Fitzgerald, 1966) was expanded
to include other aspects of experiencing suggested by C. G. Jung's
(1933) four functions. As part of an exhaustive factorial study
of the optimal personality, Coan (1974) also related openness to
early memories and general activities.

A general dimension of openness also can be seen to be implicit in a number of nonpsychoanalytic personality theories.
In particular, humanistic and growth theories like those of Maslow
and Rogers advocate a conception of positive mental health above
and beyond mere adjustment. A dimension of openness marked by
toleration for and exploration of the unfamiliar, a playful approach to ideas and problem solving, and an appreciation of experience for its own sake seems to embody this distinction.

From a somewhat different perspective, the Q1 component of
the original 16PF cluster--Liberal Thinking--suggested that open
and closed systems of thought, values, and ideology might also be
a part of the same domain. The extensive work on authoritarianism
(Adorno, Frenkel-Brunswick, Levinson, and Sanford, 1950) would
then be relevant, and links to Rokeach's (1960) conception of the
open and closed mind could be made.

Finally, a dimension of openness to experience should be of
particular importance to gerontologists. A priori it seems that
developmental changes, particularly growth, are most likely to
occur in individuals who most fully experience the events of their
lives. Adjustment to aging and to a rapidly changing culture may
be easiest for open individuals. The normative, adjustive processes of reminiscence and the life review (Butler, 1968) may be
influenced by the individual's level of openness, and the topic
of rigidity, long associated with old age in both the popular and
professional mind, can be subsumed under the general heading of
open-closedness.

There are to be sure dangers in over emphasizing the positive
aspects of openness: open individuals may be more vulnerable to
threat, more flighty in interests, more indecisive in action. But

the picture of extreme closedness--impoverished in phantasy, restricted in affect, behaviorally rigid, ideologically dogmatic,
bored by ideas and insensitive to art and beauty--is even worse.

Development and Validation of the Experience Inventory (EI)

Coan's Experience Inventory required some modification to be
of maximum utility for our purposes. Some of his scales, e.g.,
"Constructive utilization of phantasy and dreams," seemed more
appropriate to his college population and his psychoanalytically
based concepts than our own. We retained, with some item modification, three of his scales--openness to Phantasy, Esthetic experiences, and abstract or theoretical Ideas--and added three new
scales: affects or Feelings, new and different Actions, and non-
dogmatic Values. For each of the six scales fourteen items were
written.

Item factor analysis as well as consideration of balanced
item scoring were used to select final eight-item scales. Table
3 shows factor loadings and indicates the success with which the
rational scales could be separated factorially. (All 48 items
loaded in the correct direction on the first unrotated factor of
general openness.)

The Experience Inventory was intended as a replacement for
the unstable third cluster of the 16PF which, in various age groups,
was composed of scales B (Brightness), I (Tendermindedness), M
(Imaginativeness), and Q1 (Liberal Thinking). A direct test of its
validity was therefore provided by a joint factoring with those
scales which figured in the original clusters. Table 4 presents
the results of this analysis, in which rows are grouped according
to the original clusters and columns are labeled according to the
obtained factors. While 16PF Tendermindedness and EI Feelings
show their primary loadings on the Anxiety factor (in a male sample), the overall pattern of loadings confirms the idea of a general dimension of openness independent of anxiety and extraversion.

The generality of the dimension is confirmed by significant
positive intercorrelations (ranging between .10 and .30) between
all six scales. When different age cohorts are examined separ-
ately, this pattern of correlations is replicated. Because of
the age-invariant generality of openness, it is possible and useful to combine scales into a total score.

External correlates provide both convergent and discriminant
validity for the Experience Inventory. Table 5 gives correlations
of EI scales and the total score with a number of other instruments. Notable among the many significant correlations are the
positive associations with the SVIB Theoretical Interaction Style
factor and GATB Information Processing and Pattern Analysis factors.

These correlations replicate findings reported elsewhere with a measure of openness based on combining 16PF M and Q1 scales (Costa and McCrae, in press; Costa, Fozard, McCrae, and Bosse´, 1976). Evidence for the discriminant validity of the individual scales is seen in the relatively higher correlations of AVL Theoretical Values with Ideas and AVL Aesthetic Values with Esthetics, and of Traditional Family Ideology with Values. The correlation of Openness scales with a measure of Positive Emotions or affects supports the general idea that openness contributes to positive well-being. Finally, correlations with the Social Desirability Scale are small and suggest that openness is slightly negatively related to the need for approval (Crowne and Marlowe, 1964).

Table 3

Experience Inventory Items and Factor Structure

ITEMS

Phantasy

3[a] (R)[b]	I prefer not to waste my time day-dreaming.
7[a] (R)	I try to keep all my thoughts directed along realistic lines and avoid flights of fancy.
13[a] (R)	If I feel my mind starting to drift off into daydreams, I usually get busy and start concentrating on some work or activity instead.
47 (R)	I would rather tend to business in the present than dream about the future.
52[a]	I enjoy concentrating on a fantasy or daydream and exploring all its possibiliites, letting it grow and develop.
57[a]	I enjoy an active fantasy life and indulge in it fairly often.
75[a]	I can daydream for long periods of time and completely forget where I am.
83[a]	Without fantasy and daydreams, life would seem very dull and drab to me.

Esthetics

4 (R)	I like music in the background, but I seldom like to just sit and listen.
18[a]	Sometimes when I am reading poetry or looking at a work of art, I feel a strong wave of excitement that seems to affect my whole body.
29	I am often completely absorbed in music I am listening to.
30[a] (R)	Poetry has little effect on me.
55[a]	I enjoy reading poetry that emphasizes feelings and images more than story line.
71 (R)	Looking for the structure or patterns in music or literature would spoil the fun for me.
76	Great music has an endless fascination for me.
80[a]	I have had experiences which inspired me to write a poem or story, or make up a humorous tale, or paint a picture.

Feelings

2 (R)	I rarely experience strong emotions.
10	Without strong emotions, life would be uninteresting.
39 (R)	I don't like sentimental movies.
44 (R)	I rarely let myself get angry with anyone.
45 (R)	In dealing with highly charged issues, I prefer to remain detached.
53	In making decisions, I am often swayed by jealousy or pity.
68	Feelings and emotions are important guides to conduct for me.
79	I seem to be more sensitive to feelings than others are.

(Cont'd on page 130)

Age Trends in Openness

Improvements in items can be made and the task of validation is far from finished, but evidence so far justifies and encourages the continuing use of the measure. Of immediate interest to gerontologists is the question of age relations: does openness diminish with age, while rigidity sets in? Figure 1 provides a tentative answer. Analyses of variance on seven age cohorts show significant differences on four scales and the total; the pattern is roughly the same across all five measures. Instead of a decline with age cohort--either linear or accelerated--we see a saw tooth effect, peaking at age 56-60.

Table 3 (cont'd)

ITEMS

Actions

9		I enjoy trying new and foreign foods.
33	(R)	I prefer playing familiar card games to learning new ones.
35		I'm always willing to try a new way of doing an old job.
41	(R)	"Innovations" are usually more trouble than they're worth.
61		I seek out new and different places to visit.
64	(R)	On a vacation, I prefer going back to a tried and true spot.
70		I enjoy surprises even when they mean that I have to change my plans.
74		I enjoy meeting people with unusual views or background.

Ideas

6^a		I enjoy reading science fiction stories.
19	(R)	I find philosophical arguments boring.
36		If I were a scientist, I would rather create theories than collect facts.
43^a	(R)	I sometimes get annoyed by people who like to talk about very abstract, theoretical matters.
50^a		I often enjoy playing with theories or abstract ideas.
62^a		I enjoy working on "mind twister" type puzzles which require an unexpected approach to achieve solutions.
66^a	(R)	I have never been very interested in thinking up idealistic schemes to improve society.
77^a	(R)	I do not enjoy solving mathematical problems or puzzles.

Values

15	(R)	The "new morality" of permissiveness is no morality at all.
17	(R)	If a person doesn't know what he believes in by the time he's 20, there's something wrong with him.
24	(R)	I have faithfully preserved the values and teachings of my parents.
32	(R)	There is a lot of needless debate on issues like capital punishment.
51	(R)	Letting students hear radical speakers can only confuse and mislead them.
59		I feel that we can learn from the spiritual teachings of other religions, like Buddhism.
65		The different ideas of right and wrong that people in other societies have may be valid for them.
73		Laws and social policies should change to reflect the needs of a changing world.

[a]Original Coan (1972) items.

[b]Reverse-scored items marked "(R)."

(Cont'd on page 131)

If this effect is real, i.e., replicable on other samples, the question remains as to its cause. It may represent a post-mid-life spurt of psychic growth before the final decline but a more compelling argument can be made for a meaningful generational effect. Individuals living their first ten years in the stimulation of the Roaring 20's (age cohort 56-60) are as open as men 20

Table 3 (cont'd)

FACTOR STRUCTURE[c]

		Phantasy	Esthetics	Feelings	Actions	Ideas	Values
Phantasy	- 3	-61					
	- 7	-57					
	-13	-50					-33
	-47	-42				-32	
	52	64					
	57	68					
	75	64					
	83	64					
Esthetics	- 4		-41				
	18		56				
	29		63				
	-30		-60				
	55		62				
	-71		(-29)				
	76		61				
	80		50				
Feelings	- 2			-62			
	10			49	34		
	-39			-33			
	-44			-60			
	-45			-46			
	53	41		(16)			
	68			33			
	79		30	(20)			
Actions	9				43		
	-33				(-06)	-47	
	35				42		
	-41				(-27)		-36
	61				59		
	-64				-36		
	70				39		
	74				59		
Ideas	6					36	
	-19		-31			-31	-38
	36					31	
	-43					-43	-45
	50					43	
	62					60	
	-66					-33	
	-77					-57	
Values	-15						-45
	-17						-43
	-24						-55
	-32						-53
	-51						-62
	59				39		(28)
	65				33		(27)
	73				61		(16)

[c]Varimax rotation of principle components analysis, N = 844. All loadings over .30 shown; decimal points omitted. Loadings in parentheses given to show contribution of item to expected factors.

years their juniors (age cohort 30-40); men growing up in the Depression (age cohort 46-50) are as closed as their fathers (age
cohort 66+). The low openness in the oldest groups might be due
either to increasing rigidity with age, or to enculturation in the
conservative society of the early 20th Century. Further interpretation would be premature but the implications of this dimension
for the development of personality--whether in childhood or later

Table 4

Factor Analysis of Experience Inventory
and 16PF Scales[a]

	Anxiety	Extraversion	Openness
16PF			
C - Stable	-80		
L - Suspicious	56		
O - Guilt-prone	79		
Q4 - Tense	83		
A - Outgoing		59	
F - Happy-go-lucky		79	
H - Adventurous		72	
Q2 - Independent		-64	
B - Bright			34
I - Tenderminded	38		(18)
M - Imaginative			47
Q1 - Liberal thinking			(29)
Experience Inventory			
Phantasy	32		43
Esthetics			56
Feelings	42		(22)
Actions			51
Ideas			61
Values			45

[a]Varimax-rotated principle axis factor loadings, N = 409,
Total Group. All loadings above .30 shown; decimal points
omitted. Loadings in parentheses given to show contribution
of scale to expected factor.

Table 5

Correlates of Experience Inventory Scales[a]

	Phantasy	Esthetics	Feelings	Actions	Ideas	Values	Total Openness
SVIB Factors (N = 390)							
Person vs. Task	03	19***	18***	22***	08*	10*	23***
Theoretical Style	12*	35***	-03	21***	46***	32***	41***
Tough-Mindedness	-11*	-22***	05	-01	15***	-00	-05
Assertiveness	-03	-06	05	-01	-04	09*	-01
Business vs. Healing	-13**	-10*	-11*	-15***	-10*	-14**	-20***
AVL Values (N = 407)							
Theoretical	04	02	-09*	01	18***	15***	09*
Economic	-05	-25***	-04	-09	-15***	-11*	-21***
Aesthetic	06	24***	-06	07	05	11*	15*
Religious	-03	04	04	-05	-03	-19***	-07
Social	-12**	-03	05	01	-06	02	-03
Political	10*	-06	11*	06	03	08	08
GATB Factors (N = 409)							
Information Processing	06	05	04	03	18***	21***	16***
Manual Dexterity	02	-02	01	09*	04	-01	04
Pattern Analysis	17***	03	05	16***	19***	14**	21***
Positive Emotionality (N = 591)	10**	26***	31***	36***	20***	04	36***
Education (N = 409)	03	15***	-05	10*	17***	21***	18***
Warner Level SES (N = 748)	-03	-06*	-08*	-06	-12**	-14***	-13***
Traditional Family Ideology (N = 433)	-21***	-20***	-05	-23***	-25***	-61***	-42***
Crowne-Marlowe Social Desirability (N = 530)	-24***	-03	-25***	08*	-05	-19***	-18***

[a]Correlations based on all subjects for whom data available; Ns given in parentheses.

*p = .05; **p = .01; ***p = .001. Decimal points omitted.

life--are clearly worthy of further research. However interpreted, these data contribute to the growing belief that aging is not a simple process of linear decline but may involve transitional per-

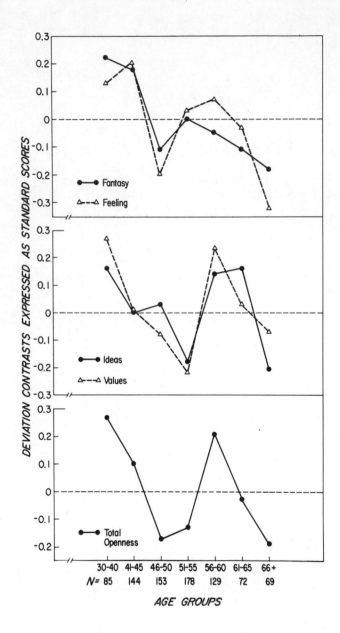

Figure 1. Cross-sectional Trends in Openness to Experience Scales and Total Scores.

iods, cycles, or phases. Some evidence on one of the better known
of these phases is presented in the next section.

OBJECTIVE ASSESSMENT OF THE MALE MID-LIFE CRISIS

In recent years, interest in the mid-life has increased among
both psychologists and laymen. Literary and clinical accounts of
a "crisis" which marks this period have formed the basis for elaborate theories of adult development. Even more attention has been
paid by a vast reading public to best-selling popularized works,
(e.g., Sheehy, 1976).

A number of theorists working from different perspectives
have converged on the idea that a period of reassessment occurs
somewhere around age 40 in which individuals must face the fact
that their lives are half over. Probable causes mentioned range
from the biological to the sociocultural (Brim, 1976). Social
psychologists like Lowenthal (Lowenthal, Thurner, and Chiriboga,
1975) note the changing roles and statuses that come with independence of children and the "empty nest." Psychologists like
Levinson (Levinson, Darrow, Klein, Levinson, and McKee, 1974)
attempt to bridge sociological and psychological perspectives in
the use of a biographical method and point to internal reorien-
tations which necessitate the creation of a new "life structure."
According to Levinson, realization that the set of early aspira-
tions making up the "Dream" may not be achieved or, when achieved,
is not as meaningful as hoped, animates the crisis at mid-life.
Psychoanalysts like Roger Gould (1975) hold that the crisis represents the final process in the dismantling of childhood illusions
of safety. More eclectic views such as those of Neugarten (1973)
point to the decline in power and vigor as well as an "increasing
interiority" which may underlie readjustments to life and pace
developmental changes in adult personality.

The mid-life crisis is hypothesized to occur when some or all
of these causes precipitate a period of self-preoccupation, inner
turmoil and external distress. As with the causes and descriptions
of the crisis, there is variation among theories as to the exact
ages or periods at which the crisis can be expected to occur.
Gould (1972) suggests it will most commonly occur between 37 and
43; Levinson puts it between 40 and 45; and Neugarten argues that
such "transitions" usually occur slowly and may occur anywhere
from the thirties to the late fifties.

Hard empirical evidence on the existence as well as incidence
of the crisis, however, is rare and often disconfirming: epidemiological indices (Kramer and Redick, 1976) fail to show an expec-
table rise or clustering of such secondary symptoms (Brim, 1976)

as divorce, alcoholism or depressive illnesses at the mid-life,
though these simple indicators may be too crude to detect a mid-
life crisis. A self-report inventory of the mid-life crisis concerns and symptoms might be sufficiently sensitive while still
providing objective, reliable measures of the condition.

Empirical Evidence

Such a scale was constructed and administered to the subjects
in the Normative Aging Study in two separate studies. The first,
an undergraduate honors project (Cooper, 1977) on 233 subjects,
was used as a pilot study. Ten content areas were identified from
a review of the literature which characterized mid-life concerns
and symptoms and items were adapted from Gould's (1972) questionnaire or newly written to measure them. The areas were (with
item examples): Marital Dissatisfaction ("My wife doesn't seem
to be there when I need her anymore"); Job Dissatisfaction ("I feel
the years I have spent at my job were meaningless and unfulfilling");
Problems with Children ("I feel uneasy that my children don't need
me anymore"); Inner Orientation ("These days I am often preoccu-
pied with myself"); Stagnation ("My life now is boring, tedious,
and unchanging"); Decline in Power and Potency ("I can't do things
as well as I used to"); Concern for Failing Parents ("It's disturbing to watch my parents grow old and weak"); Rise of the Re-
pressed ("I find myself becoming more emotional than I used to be");
and Time Push ("It hurts me to realize I will not get some things
in life I want").   Since all scales were positively intercorre-
lated, they were also combined in a total score.

An open-ended question was also included on the questionnaire
asking subjects directly if they felt they were in a mid-life
crisis, and how it felt. As an attempt at scale validation, the
scores of seven subjects whose response to this open-ended question
was clinically judged to indicate a mid-life crisis were compared
with the scores of 21 others selected at random. Seven of the ten
scales and the total score showed statistically significant differences, with clinically diagnosed subjects scoring higher in all
cases.

Item analysis of the questionnaire was used to select the best
36 items representing all ten content areas. These items were
administered to a second sample of 315 men, ranging in age from
33 to 79.

In the first study, analyses of variance showed no significant differences between mid-life and post mid-life groups. On
the premise that the crisis might be more common among middle
class than working class men, analyses were also performed with
dichotomized social class as a second classifying variable. Neither

main effects nor age-by-social class interactions were significant for any of the ten scales or total.

In the second study, using a reduced set of items, analyses were performed only on the total score. Subjects were grouped into four classifications of mid-life and post mid-life, with ages dictated by different theoretical positions on the time of onset of the crisis. Age contrasts of 35-45 vs. 46-55 (Ns = 84, 119); 37-43 vs. 44-55 (Ns = 57, 141); 35-54 vs. 55-79 (Ns = 191, 122) and 40-45 vs. 50-55 (Ns = 60, 70) showed no significant differences on total mid-life crisis scores. The last grouping did show a trend in the predicted direction (F = 3.74, p = .055), with 40-45 year olds about one-third of a standard deviation higher than post-midlifers.

In an attempt to interpret the Mid-life Crisis scale in terms of external correlates, total scores were correlated with Eysenck Personality Inventory E (Extraversion) and N (Neuroticism) scores, and with 16PF and Experience Inventory scales. Correlations of .51 with Neuroticism and -.15 with Extraversion suggested that the Mid-life Crisis scale measures, at least in part, anxiety and some introversion. While these relations are reasonable, since neurotics and introverts might be expected to be more prone to a mid-life crisis, discriminant validity requires that the scale also show relations which E and N do not. Careful examination of the correlations of Mid-life Crisis and EPI E and N scales with 16PF and EI scales (presented in Table 6) fails to provide evidence of these divergent relations. It appears that endorsement of the concerns and symptoms of the mid-life crisis indicates primarily neuroticism.

A Reinterpretation of the "Mid-life Crisis"

The lack of evidence for a distinct mid-life crisis syndrome at the expected age can be variously interpreted. The scale itself may be invalid, although it is internally consistent (coefficient alpha = .90) and does show convergent validity in its relation to neuroticism scores and to blind clinical judgments of those few subjects who reported being in a crisis. More fundamentally, the crisis may be discernible only by the probing judgment of a skilled clinician. While this possibility cannot be ruled out, it is one with which the empirically-minded psychologist will not rest easily.

However, taken at face value these results paint a very different picture of the mid-life crisis than that which is commonly accepted today. According to our data, most men do not go through a mid-life crisis at all; those who do, do so at no particular age; and the crisis itself may be nothing more than a manifestation of

long-standing instability or neuroticism. This conclusion echoes
that of Lowenthal and Chiriboga (1972) who found little evidence
for a mid-life crisis and reported that many of the problems which
their subjects had "seemed to be continuations of past difficulties rather than new problems of middle age (Troll, 1975, p. 65)."
The 16PF scales in Table 6 were administered eleven years prior
to the Mid-life Crisis scale. Thus, the correlations with C
(Stability), O (Guilt-proneness) and Q4 (Tenseness) indicate a long

Table 6

Correlates of Mid-Life Crisis Scale, EPI Extraversion and EPI Neuroticism[a]

	Mid-Life Crisis[b]	Extraversion[c]	Neuroticism[c]
16 PF Scales			
A	-11	34***	-01
B	-05	-05	-17**
C	-36***	06	-53***
E	00	31***	08
F	-04	61***	01
G	-11	-02	-09
H	-28***	45***	-26***
I	-07	-08	19**
L	22**	04	43***
M	05	-10	17**
N	-07	-04	-12*
O	32***	-16**	55***
Q1	-06	-08	-17**
Q2	-04	-31***	06
Q3	-19**	-14*	-38***
Q4	28***	-01	55***
Experience Inventory			
Phantasy	17**	06	21***
Esthetics	-10*	14**	04
Feelings	08	25***	35***
Actions	-19***	31***	-14**
Ideas	-09	08*	-04
Values	-14**	05	-13**
Total	-06	25***	09*

[a]16PF Scales administered, 1965-1967; other measures administered,
1976-1977.

[b]N = 159 for 16PF Scales; 288 for Experience Inventory Scale.

[c]Eysenck Personality Inventory Scales; N = 241 for 16PF Scales;
448 for Experience Inventory Scale. Decimal points omitted.

*p = .05; **p = .01; ***p = .001

history of neurotic difficulties for current mid-life crisis suf-
ferers and diminish the possibility that a life-crisis has caused
the symptoms measured in the contemporaneous Eysenck measure.
Life-long anxiety may manifest itself in adolescent identity
crises, alcoholism, or classical phobias and obsessions; among
men in the middle part of their lives, it is liable to manifest
itself as complaints about marriage, family, impending death, and
career dissatisfaction. Rather than a period of restructuring
one's life, the "mid-life crisis" may simply be a period of generalized complaint by individuals less able to cope with the life
they have.

The creative crisis which spurs the imagination of developmentalists, which stems from a failure of existential meaning in
life, which marks the careers of such figures as Whitman and
Gauguin--this crisis seems to be a rarity, restricted perhaps to
the tiny portion of mankind characterized by Maslow as "self-
actualizing." If this is true, both theory and the clinical practice which grows from it have been led astray. To assume that
chronic neurotic complaining is acute existential crisis or a
normative developmental transition must lead to inappropriate therapy. Rather than viewing it as an expression of a life-span developmental process, which should be supported and encouraged, the
clinician might better consider the application of standard clinical methods.

THE ROLE OF OBJECTIVE ASSESSMENT IN CLINICAL GERONTOLOGY

The evidence cited in this chapter has hopefully made it clear
that objective, and in particular, self-report methods can make
significant contributions to an understanding of adult personality.
Objectively measured personality traits--at least the broad dimensions of anxiety, openness, and extraversion--are meaningful and
stable dimensions of individual differences which can be measured
by standard tests applicable to adults through the eighth decade.

The uses of objective measures documented in this chapter
have been chiefly limited to research purposes in which group differences are examined. Such research has, of course, tangible
benefits for the clinician. It can provide a rough sketch of normal personality which can be used to separate pathological changes
from normal developmental processes and it can provide "hard" evidence for the evaluation of developmental concepts like the "mid-
life crisis."

The use of objective measures in the clinical assessment of
individuals, however, is another and more complex issue, covered
only briefly here. The clinician as intelligent consumer of

standardized tests must bear several points in mind, beginning
with the suitability of personality testing at all: sensory, cognitive, and motivational problems may invalidate any objective
test, and pose particular problems for elderly patients, especially
when organicity or senility may be present. In the choice of an
objective instrument, the clinician should consider all of its
psychometric properties carefully. For example, measures with
impressive evidence of validity when used for research on group
comparisons may have errors of measurement so large that they have
severely limited value in the assessment or diagnosis of individuals. Similarly, the availability of appropriate norms and
clinically meaningful cut-off points must be considered in the
choice of a particular test. As Lawton's review (Lawton, Whelihan, and Belsky, in press) points out, the evidence for making
these evaluations is still scarce. Consequently, objective tests
should be used with caution and form only a part of the basis of
clinical judgments.

From a more optimistic point of view, our review of the literature and our own data support the position that objective
measures can be used as a meaningful adjunct to other forms of
clinical assessment. The stability of personality throughout
adulthood suggests that tests and norms developed on young adults
may be used at least until more adequate tests or procedures become available for older adults. Clinicians in general and clinical gerontologists in particular might also profitably broaden the
scope of their assessments beyond pathology-based conceptions and
measures like the MMPI. Other dimensions of personality, especially experiential openness, may be equally valuable in assessing
the quality of functioning in the elderly.

REFERENCES

Aaronson, B. S. Age and sex influences on MMPI profile peak dis-
    tributions on an abnormal population. Journal of Consulting
    Psychology, 1958, 22, 203-206.
Adorno, T. W., Frenkel-Brunswick, E., Levinson, D. J., and San-
    ford, R. N. The authoritarian personality. New York: Harper,
    1950.
Bell, B., Rose, C. L., and Damon, A. The Normative Aging Study:
    An interdisciplinary and longitudinal study of health and
    aging. Aging and Human Development, 1972, 3, 5-17.
Brim, O. G., Jr. Theories of the male mid-life crisis. The
    Counseling Psychologist, 1976, 6, 2-9.
Brozek, J. Personality changes with age: An item analysis of
    the MMPI. Journal of Gerontology, 1955, 10, 194-206.
Butler, R. The life review: An interpretation of reminiscence in
    the aged. In B. L. Neugarten (Ed.), Middle age and aging: A

reader in social psychology. Chicago: University of Chicago Press, 1968.

Calden, G., and Hokanson, J. E. The influence of age on MMPI responses. Journal of Clinical Psychology, 1959, 15, 194-195.

Cattell, R. B. The descriptions and measurement of personality. Yonkers, N.Y.: World Book, 1946.

Cattell, R. B., Eber, H. W., and Tatsuoka, M. M. Handbook for the Sixteen Personality Factor Questionnaire. Champaign, Ill.: Institute for Personality and Ability Testing, 1970.

Chown, S. M. Age and the rigidities. Journal of Gerontology, 1961, 16, 353-362.

Chown, S. M. Personality and aging. In K. W. Schaie (Ed.), Theory and methods of research on aging. Morgantown, W.V.: West Virginia University Press, 1968.

Coan, R. W. Measurable components of openness to experience. Journal of Consulting and Clinical Psychology, 1972, 39, 346.

Coan, R. W. The optimal personality: An empirical and theoretical analysis. New York: Columbia University, 1974.

Cooper, M. W. An empirical investigation of the male midlife period: A descriptive, cohort study. Unpublished undergraduate honors thesis, University of Massachusetts at Boston, 1977.

Costa, P. T., Jr., and McCrae, R. R. Age differences in personality structure: A cluster analytic approach. Journal of Gerontology, 1976, 31, 564-570.

Costa, P. T., Jr., and McCrae, R. R. Age differences in personality structure revisited: Studies in validity, stability, and change. Aging and Human Development, in press.

Costa, P. T., Jr., Fozard, J. L., McCrae, R. R., and Bosse´, R. Relations of age and personality dimensions to cognitive ability factors. Journal of Gerontology, 1976, 31, 663-669.

Costa, P. T., Jr., Fozard, J. L., and McCrae, R. R. Personological interpretation of factors from the Strong Vocational Interest Blank scales. Journal of Vocational Behavior, 1977, 10, 231-243.

Crowne, D., and Marlowe, D. The approval motive. New York: Wiley, 1964.

Douglas, K., and Arenberg, D. Age changes, cohort differences, and cultural change on the Guilford-Zimmerman Temperament Survey. Unpublished paper, 1977.

Eysenck, H. J., and Eysenck, S. B. G. Manual for the Eysenck Personality Inventory. San Diego: Educational and Industrial Testing Service, 1968.

Fiske, D. W. Measuring the concepts of personality. Chicago: Aldine, 1971.

Fitzgerald, E. T. Measurement of openness to experience: A study of regression in the service of the ego. Journal of Personality and Social Psychology, 1966, 4, 655-663.

Gould, R. L. The phases of adult life: A study in developmental psychology. American Journal of Psychiatry, 1972, 29, 521-531.

Gould, R. L. Adult life stages: Growth towards self-tolerance.
    Psychology Today, 1975, 8, 74–78.
Guilford, J. S., Zimmerman, W. S., and Guilford, J. P. The Guil-
    ford-Zimmerman Temperament Survey handbook. San Diego:
    EdITS Publishers, 1976.
Gutmann, D. L. An exploration of ego configuration in middle and
    later life. In B. L. Neugarten (Ed.), Personality in middle
    and later life. New York: Atherton Press, 1964.
Gynther, M. D., and Shimkunas, A. M. Age, intelligence and MMPI
    F scores. Journal of Consulting Psychology, 1965, 29,
    383–388.
Hardyck, C. D. Sex differences in personality changes with age.
    Journal of Gerontology, 1964, 19, 78.
Hogan, R., DeSoto, C. B., and Solano, C. Traits, tests, and per-
    sonality research. American Psychologist, 1977, 32, 255–264.
Hundleby, J. D. The measurement of personality by objective tests.
    In P. Kline (Ed.), New approaches in psychological measure-
    ment. London: John Wiley, 1973.
Jung, C. G. Psychological types. New York: Harcourt, 1933.
Kramer, M., and Redick, R. W. Epidemiological indices in the
    middle years. Unpublished paper, cited in O. G. Brim, Jr.,
    Theories of the male mid-life crisis. The Counseling Psychol-
    ogist, 1976, 6, 2–9.
Lawton, M. P., Whelihan, W. M., and Belsky, J. K. Personality
    tests and their uses with older adults. In J. Birren (Ed.),
    Handbook of mental health and aging. New York: Prentice Hall,
    in press.
Levinson, D. J., Darrow, C. M., Klein, E. B., Levinson, M. H., and
    McKee, B. The psychosocial development of men in early adult-
    hood and the mid-life transition. In D. F. Ricks, A. Thomas,
    and M. Roff (Eds.), Life history research in psychopathology,
    Vol. 3. Minneapolis: University of Minnesota Press, 1974.
Levinson, D. J., and Huffman, P. Traditional family ideology and
    its relation to personality. Journal of Personality, 1955,
    23, 251–273.
Lowenthal, M., and Chiriboga, D. Transition to the empty nest.
    Archives of General Psychiatry, 1972, 26, 8–14.
Lowenthal, M. F., Thurner, M., and Chiriboga, D. Four stages of
    life. San Francisco: Jossey-Bass, 1975.
Mischel, W. Personality and assessment. New York: Wiley, 1968.
Neugarten, B. L. Personality change in late life: A developmen-
    tal perspective. In C. Eisdorfer and M. P. Lawton (Eds.),
    The psychology of adult development and aging. Washington,
    D. C.: American Psychological Association, 1973.
Neugarten, B. L. Personality and aging. In J. E. Birren and
    K. W. Schaie (Eds.), Handbook of the psychology of aging.
    New York: Van Nostrand Reinhold, 1977.
Rokeach, M. The open and closed mind. New York: Basic Books,
    1960.
Schaie, K. W. A general model for the study of developmental prob-

lems. _Psychological Bulletin_, 1965, _64_, 92-107.

Schaie, K. W., and Marquette, B. Personality in maturity and old
    age. In R. M. Dreger (Ed.), _Multivariate personality re-
    search: Contributions to the understanding of personality
    in honor of Raymond B. Cattell_. Baton Rouge, LA: Claitor's
    Publishing, 1972.

Schein, V. E. Personality dimensions and needs. In M. W. Riley
    and A. Foner (Eds.), _Aging and society_. Vol. 1. New York:
    Russell Sage, 1968.

Sheey, G. _Passages: Predictable crises of adult life_. New York:
    Dutton, 1976.

Swenson, W. M. Structured personality testing in the aged: A
    MMPI study of the gerontic population. _Journal of Clinical
    Psychology_, 1961, _16_, 49-52.

Troll, L. E. _Early and middle adulthood_. Monterey, CA: Brooks/
    Cole, 1975.

THE USE OF PROJECTIVE TECHNIQUES IN PERSONALITY

ASSESSMENT OF THE AGED

Boaz Kahana

Oakland University

The classical problem confronting personality assessment in the elderly as well as in other groups has been that of surveying the gestalt of personality using indices which are both valid and reliable and yet which allow an overall understanding of the person (Chown, 1967). Hacker, Gaitz, and Hacker (1972, p. 94) argue that a humanistic view of mental health dictates a reevaluation of our orientation to diagnostic assessment. "In the pressure to quantify certain aspects of human behavior for scientific investigation, the essential human or spiritual qualities of the person are de-emphasized, and at times forgotten." Behavioral scientists increasingly have been using standardized, closed-ended indices of mental health or illness. Data obtained through such measures are easily scored, processed, and quantified. Research on older persons has shown, however, that such indices may not capture the complexities of behavior and may even encourage a dehumanizing orientation toward older mental patients. Clinical assessment of the aged must go beyond an attempt to characterize mental impairment and illness and serve to uncover the potential of the aging individual (Oberleder, 1967).

The direction typically taken in gerontology in seeking techniques for diagnostic evaluation of the aged has been that of "functional" assessments (Howell, 1968; Gaitz and Baer, 1970). Such efforts have been geared to determine whether older persons can perform the tasks necessary for existence in their environment. Efforts have been made to devise a comprehensive tool for functional assessment, but such attempts have thus far not yielded a single widely accepted index or instrument. Typical indices of functional assessment include only minimal information on psychological functioning. When attention is directed to psychological

functioning it is usually limited to mental status and cognitive
impairment and does not focus on the dynamic aspects of personality.

Schaie and Schaie (1977) in their recent review of clinical
assessment with the aged argue for the proper use of projective
techniques for the testing of clinical hypotheses along lines of
the hypothetico-deductive experiment with the client as a sample
with a "N of one." They outline three major objectives for such
assessment: (1) diagnosis of psychopathology; (2) determination
of baseline behaviors which permit comparisons following behavioral
intervention; and (3) assessment of adjustment to role changes.

With increasing recognition of the rights of older persons to
self-improvement, therapeutic growth, and self-actualization,
projective tests may be viewed as increasingly useful and needed
assessment tools and as adjuncts to intervention. The use of psychological tests to consider individual personality differences
among the aged also reflects a movement away from stereotyping all
aged individuals as "basket cases" and toward recognizing the
great range of characteristics among older persons. In contrast to
the simplistic attempts to disregard "wounded aspects of the personality" which typify some of the cognitive therapies with the
aged, the use of personality tests is congruent with a humanistic,
socioemotional approach. Prior to therapy, it is useful to gain
insights into intrapsychic strengths as well as problems.

Suitability of Projectives for Use with the Aged

Most commonly used projective tests have been designed in a
"nondevelopmental pathology-oriented context" (Schaie and Schaie,
1977). Norms for some of these tests have been collected later
for the aged. Yet even in these cases, scoring systems or techniques of administration may not be suitable to the aged, or may
have different meanings. Occasionally, special procedures for use
with the aged were developed in a research context (Gutmann, 1969;
Birren, Butler, Greenhouse, Sokoloff, and Yarrow, 1963). However,
these research uses have seldom been translated to, or utilized in,
clinical situations.

Where quantitative norms for performance are not available,
clinicians may rely on their own clinical experiences in interpreting test data. Unfortunately most clinicians lack even these
qualitative yardsticks in interpreting test responses for the aged.
Furthermore, clinical training programs seldom include experience
with older clients, thereby stripping the would-be clinician of
reliance on formal training or supervision. Without such clinical
acumen in work with the aged, the clinician vacillates between

attributing most decrements and problems in test performance to
normal developmental changes or reading extensive pathology into
test responses of relatively intact aged.

The Case for the Use of Projectives

In assessing the value of projective tests for use with the
aged, one must confront the issue which has been heatedly debated
in psychology about the use of projective testing in general
(Klopfer and Taulbee, 1976).

In understanding criticisms regarding poor or unknown reliability and validity of projective tests, it is useful to note that
most of these measures represent a technique rather than a "test."
Reliability would therefore have to be viewed in more general terms
such as similarity within an individual, of themes, conflicts, and
coping strategies over time or across tasks. The latter approach
is problematic since different parts of a test have a different
stimulus pull. Nevertheless, one often observes within the same
subject a fair degree of regularity across tasks in types of conflicts and coping styles portrayed. Projective tests portray unconscious or preconscious fantasies, conflicts and coping mechanisms. It is therefore difficult to validate data derived from
these measures against behavioral indices or conscious self-reports
as found in personality questionnaires (Guilford, 1950). The
analysis of other projective tests and of dreams might provide
meaningful validation of these instruments. Projective tests have
been criticized in numerous methodological reviews for lack of
sufficient reliability and validity and especially for their inability to predict behaviors. Yet, the Annual Review of Psychology
(Klopfer and Taulbee, 1976) lists over 500 articles published on
the subject during 1971-1974 and clinicians everywhere continue to
rely on projectives as major tools for assessing personality
dynamics. The Rorschach, the TAT, the Bender-Gestalt, and the
Machover Draw-A-Person Test have largely retained their ranking as
the most frequently used projective tests, with the Rorschach and
TAT referenced 4,202 and 1,334 times respectively in the most recent Mental Measurement Yearbook (Buros, 1972).

The original popularity of projectives coincided with the general ascendancy of psychoanalytic theory. The type of questions
posed by these tests fitted well with that theoretical framework
and with the psychology of individual differences. Recent changes
towards a more behavioristic orientation have stressed universalis-
tic principles and de-emphasized personality differences. When
therapy is centered around such techniques as behavior modification,
the understanding of personality dynamics through projective tests
is not of great interest. There has been a developing recognition,

however, that no one technique, no matter how useful, can serve to
help all patients. Therapeutic intervention must be matched to
particular needs of patients (Klopfer and Taulbee, 1976). For
aged patients, tailor-made treatment programs are especially important. We cannot effectively use simple formulas provided by reality orientation and remotivation to help relatively intact older
persons cope with the special stresses which they face. Self-
concepts, values, and intrapsychic problems need to be evaluated in
order to intelligently work toward therapeutic change. Projective
tests, in addition to tapping the unconscious, can tell us about
self-concept, values, and goals often far better than other assessment techniques. It has also been pointed out that using a mental
health rather than a mental illness paradigm, projective tests can
readily portray creative capacities, hidden resources, and human
potentials of the individual. Projective and clinical tests also
present a potentially useful approach to measurement of adaptation.
In a longitudinal study of adaptation to institutional life, Kahana
and Kiyak (1976) used the Draw-A-Person, Bender-Gestalt, and Sentence Completion tests. These measures allow a dual opportunity to
assess coping behavior by giving the respondents a set of standardized, "minisituations" in which subjects are observed and evaluated.
Such behaviors as corrections of drawings, page rotations, requests
for more information in handling the test, and latencies in sentence completion responses provide behavioral clues to adaptation.
A second opportunity to assess coping is through test content.
These tests present information about intrapsychic and even cognitive aspects of coping.

Assessment of Mental Health Service Needs

Older people are reluctant to acknowledge service needs or to
request professional assistance for their problems. This has been
the case even where there is no social stigma associated with the
problem. In the case of problems related to mental health, potential stigma makes the acknowledgement of problems especially difficult. Estimates of needs for mental health services for elderly
people in the community are therefore especially difficult to obtain. In surveys of community samples, needs for services are
often inferred from characteristics of the group studied. Thus,
prevalence rates for certain conditions of functional impairment,
chronic conditions, or nutritional deficiency may be applied to
determine the level of need.

Another approach to the study of needs for service is to ask
potential target groups about their self-perceived needs and preferences. While this approach has certain advantages in that it aims
to get at the criterion rather than making inferences about it, several problems still remain. Aged subjects deny even pressing needs

because of the social stigma attached to being in need and dependent on others. Self-diagnosis in many instances may be difficult
because of the complex set of needs confronting the functionally
impaired and because the respondent may be too emotionally involved
with his own situation.

The third method for assessment of needs relies upon professional or expert judgment about the nature of the need. It is in
this latter category that projective tests may serve as an extremely valuable adjunct to more global clinical interviews and
other assessment techniques.

Before providing a detailed review and discussion of specific
projective techniques, a number of methodological issues relevant
to clinical and research use of these tests will be briefly reviewed.

Issues in Administration of Projective Tests

The aged may be perceived as frustrating the examiner attempting to use projectives in a number of ways. It is sometimes difficult to test an elderly person without some modification of testing procedure in order to account for the existence of perceptual
and/or motor difficulties. Assessment and test interpretation is
problematic because of scarce or inadequate norms for the aged,
or because the meaning of test results could be cohort specific
rather than reflective of a decline in functioning. If the examiner or therapist is not an older person himself, he may find
it difficult to empathize with the patient or to fully understand
the meaning of the patient's life events and phenomenology. Resulting frustrations may cause a variety of defenses and hostile
reactions on the part of the examiner or therapist.

In research use of projectives, problems in obtaining test
results from the aged come from two major sources: (1) The "vol-
unteer" population of elderly tested for a research project may
be reluctant "volunteers." The elderly may not mind responding
to survey-type research questions, but they may raise their eye-
brows when the researcher becomes an examiner and probes into
their psyches. Oftentimes they will ask, "What are you trying to
find out from that test?" or "Is this test supposed to tell you
whether or not I'm crazy?" (2) The level of skill and sophistication of the user of clinical tests is another source of difficulty.
The researcher who includes projective tests in the interview
schedule is oftentimes a graduate student in the area of adult
development, sometimes with some psychological clinical background.
More often than not, he or she has had very limited clinical
training, if any, in the use of projective tests and becomes very

defensive when challenged by the elderly who are not very eager to participate in lengthy studies. In contrast, an elderly individual who comes for psychological help portrays a very different orientation toward being examined. This person needs help and does not question the administration of projective tests. Furthermore, the clinical psychologist is experienced in testing a variety of difficult patients (manics, depressives, schizophrenics).

In summary, it is not at all clear that the aged are a more difficult group to examine than other clinical psychiatric groups, although methods of administering psychological tests may need to be altered to compensate for central nervous system changes and adjusted to special physical and emotional needs of the aged. As with subjects of all age groups, it is important to reduce anxiety as much as possible by establishing a relaxed atmosphere in testing aged persons who might already be suffering from a lack of self-esteem. Older persons often need reassurance in order to overcome a cautious and often self-deprecating attitude in response to test demands. Such reassurance comes across as genuine if respondents can relate their own objectives (e.g., seeking treatment) to the test situation (Klopfer, 1974). Greater variability in functioning of elderly people may make testing the limits a necessary and useful approach (Schaie and Schaie, 1977). The clinician must also consider the extent to which the observed behavior provides an accurate example of the older person's behavior in real life situations (Gaitz, 1973). This is an especially important consideration since older persons are known to be less test-wise and find the process of test taking more stressful than do younger groups.

Practice Effects

In the course of a study of the effects of psychoactive drugs on elderly psychiatric patients, Lehmann, Ban, and Kral (1968) examined practice effects in psychological test performance. The study was conducted with 107 psychiatric in-patients and a very small sample of young and old out-patient normal controls. It was found that the word association test was the only projective measure which demonstrated practice effects in all groups of respondents. The number of tests with positive practice effects was also found to increase with the presence of psychopathology, institutionalization, and age.

The Effects of Socioeconomic Status, Education, and Illness

Very little research has been done on the effects of these

background variables on the projective test performance of the
aged. Yet, it is important to note that people with varying ethnic, cultural, and social class backgrounds differ in their interpretation of desirable and stressful events and in their behavioral
as well as intrapsychic responses to such events (Dohrenwend,
1966). The role of education in determining test performance of
the aged has been stressed by Granick and Friedman (1967). They
argue that the negative correlation between education and age is
greatly responsible for the decline in test performance with age.
They found a decline in performance on 27 of 33 tests when education was not controlled. When education was controlled for, only
19 tests showed significant decline with age.

Even while recognizing the importance of personal demographic
background on behavior and test performance it may be argued that
projectives present a relatively culture free method of personality
assessment (Henry, 1956). Their successful use in many cross-
cultural studies (Gutmann, 1969) and with populations with low educational attainment makes them more suited for use in older populations whose formal education is typically limited and who are
over-represented among the poor and the ethnic.

The danger of interpreting age differences in personality and
other measures as reflecting changes over time frequently has been
emphasized in recent gerontological literature (Schaie, 1967;
Chown, 1967). Such differences in personality between different
age groups may in fact reflect cultural or background differences
between the generations. It is usually also very difficult to determine and separate the effects of illness and those of aging in
the assessment process. As Gaitz and Baer (1970, p. 47) stated:
"The functional status of the various systems is typically inter-
dependent: physical and psychological functions are often intri-
cately interrelated and are frequently complicated by social class
factors and cultural attitudes."

Intercorrelation of Various Personality
Measures Among the Aged

Low or moderate intercorrelations have been observed among
various personality tests in the general psychological literature.
Research focusing on personality of elderly subjects through the
use of objective or projective tests shows no exception to this
problem. A few illustrations underscore this point. To answer
the question: "Do self-reported coping styles correlate with projective techniques in measuring coping?" intercorrelations among
various indices of coping were examined by Kahana and Kiyak (1976).
Results of intercorrelations indicate that coping behaviors as
demonstrated on the Bender-Gestalt and Draw-A-Person tests tap the

same dimensions, but that the coping scale and the Sentence Completion task are not highly correlated with coping style as measured by the Bender and Draw-A-Person tests. This general pattern was observed across different dimensions of coping, i.e., instrumental, affective, and escape mechanisms. Hacker, Gaitz, and Hacker (1970) in a study of 27 aged mental patients found no significant correlations between the affect balance scale (Bradburn, 1964), the life satisfaction index (Neugarten, Havighurst, and Tobin, 1961), and any of the several psychiatric factor scores of mental illness in the mental status schedule. The authors attribute this lack of convergence to differential conceptualizations and multidimensionality underlying the various indices of mental health and mental illness. This suggests that assumptions and value orientations embedded in various indices of mental health and illness need to be recognized in order to meaningfully interpret test results.

Approach and Scope of this Chapter

Typical reviews of clinical assessment with the aged have pointed to age differences in performance and sensitized clinicians as well as researchers to the many methodological problems inherent in the use of projective tests (Chown, 1967; Schaie and Schaie, 1977). Little practically useful information, however, has been made available to potential users of such tests. In the present chapter, an attempt will be made to provide some practical pointers based on the author's research and clinical use of the projective tests. One of the major problems limiting the usefulness of information about the value of various projectives with the aged is the lack of meaningful conceptual and operational definitions about domains of personality and mental health being considered. The most prevalent conceptual framework noted in previous reviews (Chown, 1967) has been that of the disengagement-activity theories. Yet the limitations of this approach for a determination of either subjective or objective well-being or mental health have been well documented (Hochchild, 1975). In seeking to understand the personality processes, rather than outcomes, traditional conceptualizations of adjustment also appear limited (Britton, 1963). Even those recent papers which exhort researchers and clinicians to define the questions asked of diagnostic tests proceed to review them in a traditional global manner (Schaie and Schaie, 1977).

The present review will utilize a conceptual framework that has been found by this author to be useful for understanding environmental mastery on important components of mental health in older persons. The model used is not a traditional pathology model but rather an interactionist-developmental one based on

Piaget's notions of assimilation and accommodation and on Lewin's
(1951) field theoretical model. The use of such a framework provides a useful focus in considering the utility of selected projective tests with the aged. A focus on adaptive strategies permits
considerations of potential responses of the aged to stressful life
events and allows an examination of issues related to stability
versus change in the personality of older adults, especially in
the face of environmental changes. Ultimately, mental health may
be viewed as an ability to successfully cope with or negotiate
with one's environment.

In addition to consideration of adaptive strategies, examination of a number of personality traits appears to be especially
salient and amenable to clinical assessment by projective tests.
This includes characteristics where age-related changes are especially likely and environmental influences are noteworthy, i.e.,
variables such as affective expression, dependency, and rigidity
versus flexibility.

In considering individual tests, the following general issues
will be covered wherever information is available: (1) value and
background of the tests; (2) issues in administration and interpretation of the test with the aged; (3) reliability and validity;
(4) norms and age differences; (5) differences between community
and institutionalized aged; (6) effects of intervention; and (7)
modifications of the test or alternative forms. The various uses
of the test as a measure of adaptation or adjustment, of cognitive
functioning or impairment, or of other aspects of psychopathology
will also be discussed.

THE THEMATIC APPERCEPTION TEST (TAT)

The TAT is one of the three most widely used projective tests,
both clinically in personality assessment and in diverse social
psychological studies of personality (Klopfer, 1974). The TAT
provides a flexible instrument which can be used in diverse theoretical contexts. There has been a wide array of scoring systems
proposed and utilized for quantifying TAT responses. In addition,
content of responses to various TAT cards may also be qualitatively analyzed especially for content. Although some clinicians
stress the suitability of TAT for assessing the content of personality rather than assessing personality structure, Henry (1956)
aptly argues that such a dichotomy does not accurately fit the
situation. In his classic book, The Analysis of Fantasy (1956),
he presents a useful framework for determining both personality
structure and content through the use of the TAT. Information
provided by the TAT may reveal both idiosyncratic elements of the

individual personality and psychological features which are common
to groups. This flexibility has resulted in the development of
diverse stimulus cards, many adaptations related to use in different cultures and with different populations, and successful use in
numerous cross cultural studies of personality (Gutmann, 1969).

The TAT has been employed in developmental studies of adult
personality (Chown, 1967) but little has been written about the
clinical use of the TAT with older persons (Klopfer, 1974). It is
noteworthy that none of the previous reviews of the use of the TAT
with the aged (Chown, 1967; Schaie and Schaie, 1977) specifically
dealt with the use of the TAT either for diagnostic or baseline
purposes or for assessment of therapeutic change. Clinicians
typically administer selected TAT cards as part of a diagnostic
test battery, but there has been little uniformity in the choice
of TAT cards administered. An abbreviated series of cards (Hartman, 1970) has been recommended which has been found to elicit the
highest thematic production in comparison with the complete set.

Verbal productivity of older people on the TAT is often quite
limited. Relatively intact older persons often provide only descriptions of the cards rather than complete stories with a beginning, a middle, and end (Kahana and Kahana, 1968). This reflects a certain amount of concreteness in responses to the TAT on
the part of these older adults. There have been a number of attempts to develop special TAT cards or instruments which depict
themes relevant to the aged. Wolk and Wolk's (1970b) Geriatric
Apperception Test and Bellak's (1975) Senior Apperception Test
both represent efforts in this direction.

There also has been little uniformity in scoring systems utilized in diverse studies using the TAT and special norms for the
aged do not exist. In clinical use qualitative judgments are
typically made based on the examiner's experiences with the test
rather than on formal comparisons with existing norms.

Research on the use of the TAT with older persons may be
divided into a number of categories. Some studies have sought to
detect age-related changes in personality in normal populations
(Chown, 1967). One group of studies utilized the TAT to describe
and depict personality dynamics such as adaptation (Neugarten and
Gutmann, 1968) and adjustment (Britton, 1963). Gutmann (1964) has
used the TAT in this manner in a series of cross-cultural studies.
The TAT has also been utilized to predict psychological deterioration after institutionalization (Lieberman and Caplan, 1970).
There have also been some attempts to utilize the TAT to measure
the impact of intervention programs (Kahana and Kahana, 1970).

The TAT as a Measure of Adjustment

The thematic apperception technique depends on projecting one's own personal meaning onto presumably neutral stimuli. Britton (1963) utilized three selected TAT cards (6BM, 7BM, and 10) with a normal community-living rural aged population. The older hero described by respondents was rated in terms of his adequacy in initiating action, his self-confidence, his intellectual functioning and interpersonal acceptance. (Interjudge agreements on total ratings of 23 subjects were, respectively, .88, .93, and .98.) TAT adjustment scores derived in this manner were found to be significantly correlated with evaluations of the individual by community members. Older persons regarded well by others portrayed self-assurance and competence on the TAT. This finding may be interpreted as validating the use of TAT as a measurement of adjustment among the aged.

Lieberman and Caplan (1970) found the TAT sensitive to processes of decline which occurred antecedent to the death of the institutionalized aged. The TAT was also found to be one of the best indices of attitudes toward death in a study by Shrut (1958). Time orientation has been assessed using past, present, and future oriented stories on the TAT in a number of studies of older persons (Postema, 1970; Fink, 1957). Institutionalized aged were found to be less concerned with the future than community aged (Fink, 1957). Older persons (aged 61-76) showed greater concern with the past than did the younger ones (aged 50-60).

Age Differences in TAT Responses

Neugarten and Gutmann's (1968) classic study of personality changes with aging was based on TAT responses of normal older adults. Results of this study suggested that older persons see the environment as more difficult to master and attribute less importance to affective needs than do middle aged adults. Rosen and Neugarten (1960) used the TAT in an earlier study which also suggested decline in affect and energy with old age. In a national sample which included an age span of 21 to 65, Veroff, Atkinson, Feld, and Gurin (1960) found need-achievement to decrease with age among men and need-affiliation and need for power to decrease with age among women.

Gutmann (1964) developed a useful set of adaptive typologies based on a series of cross-cultural studies which utilized the TAT. He classified responses of older subjects as reflecting (1) active mastery; (2) achievement doubt; (3) adaptive retreat; (4) fixed conformity; (5) defective coping. He noted a general retreat from environmental mastery with increasing age. Gutmann's

modification of the TAT serves as a measure of three styles of
coping with the environment: active, passive, and magical. The
active style refers to an assertive active approach of coping with
the environment. The passive category refers to a withdrawal from
active engagement with the external environment and from aggressiveness and self-assertiveness. The magical includes stories
in which there are gross misinterpretations and distortions of
stimuli which reflect misperception of the environment and ego
regression. Although this measure was originally interpreted as
an index of psychological disengagement, it also provides a measure of environmental coping styles and adaptational abilities
(Gutmann, 1975).

     This method has been utilized by Gutmann (1975) in a series
of research studies since the scoring scheme may be applied to any
TAT story. Each story is categorized as portraying either an
active, passive, or magical mastery mode. Currently there is no
combined score for an aggregate of cards. Depending on the study,
Gutmann uses a combination of different TAT cards which includes
some of the original Murray test cards (1, 2, 6, 7BM, 10, and 17BM).
He also uses specifically constructed cards to suit the cultural
group to which it is administered (e.g., "desert scene," "a family
scene," and "old man scene"). Special cards have been administered
to various American Indian groups and to a group of Druze tribesmen
in Israel and Syria. In a review of data collected from Kansas
City, Lowland and Highland Maya and Navajo aged Indians, Gutmann
(1975) presents detailed norms on three TAT cards scored for coping
styles. Data from interview material reveals that younger men
among the Navajo and Maya groups are happiest when producing work
or being instrumental (active mastery), whereas older men in these
groups are happier when visited by relatives, hearing pleasant
music, and seeing a pretty scene (passive mastery). Longitudinal
studies of Navajo and Druze subjects show that changes within individuals over a five year period replicate those found between age
cohorts in both of these societies, i.e., there is a shift from the
active to the passive coping mode in the same individual (Gutmann, 1975).

     Only a few studies have utilized the TAT to assess effects of
therapeutic intervention with the aged. In an experimental study
of the effects of age integration versus age segregation in a hospital environment, Kahana and Kahana (1970) found greater affective expression and interaction portrayed in TAT responses of
patients placed in age integrated wards than for those in age segre-
gated settings. These findings correlated well with improvements
measured by other procedures.

Alternate Forms to the TAT: The Geriatric
Apperception Test (GAT) and the Senior
Apperception Test (SAT)

    Several adaptations of the TAT have been specifically designed
for use with the elderly. The major revisions include the GAT
(Wolk, 1972) and the SAT (Bellak, 1975). The GAT was developed to
compensate for weakness in existing apperceptive instruments which
depicted younger individuals and activities. The GAT was therefore
aimed at eliciting themes and concerns unique to the aged, such as
loss of attractiveness and sexuality, physical limitations, and
family problems. The GAT utilizes 14 specially constructed pictures, each one depicting an older person in a situation common to
the aged. Instructions for administration and interpretation are
parallel to those of the TAT.

    The SAT is a projective test similar to Murray's Thematic
Apperception Test which provides an analysis of fantasy, of needs
and press, of psychodynamic conflicts, and of modes of coping and
defense. Bellak (1975) describes the SAT as a technique, rather
than a measure. It stems from the idiographic method of personality study and does not rely very heavily on normative data. The
clinician uses the data derived from this technique to develop a
portrait of complex interrelationships. The test consists of 16
pictures depicting elderly people alone or with others in varied
situations. The stimulus cards present diverse interpersonal
situations depicting change and requiring subsequent adjustment
(e.g., moving out of one's home) and depicting conflict and negative affect. In addition, deprivations of old age are portrayed
(e.g., economic deprivation). Stimulus cards reflect significant
potential problems of the aged as viewed by the author based on
his clinical work and his review of the literature. The situations
portrayed are sufficiently broad and vague to encourage the subject
to project his or her own fantasies, anxieties, and conflicts into
the stories and to reveal the individual's style of coping and
resolving problems. Bellak attempted to use pictures which have
sufficient stimulus pull to elicit stories of substantial length
and also included pictures which would elicit popular themes so
that deviations could be noted when they occur.

    Bellak's data on the SAT are based on responses from 100 sub-
jects (46 males, 54 females, aged 65-84) representing a variety of
socioeconomic classes. In addition, four published pilot studies
of elderly people aged 60-85 with an N of 15-20 each are cited in
his book. These are compared with similar subsamples of college
students. Overall, the unpublished pilot studies reported by
Bellak generally support the usefulness of the SAT with aged populations and its sensitivity to age differences.

Research thus far has not clearly demonstrated that specially designed cards elicit significantly more detailed or revealing responses from the aged than the traditional TAT cards. Traxler, Swiener, and Rogers (1974) studied the usefulness of the GAT in assessing personality dynamics of community and institutionalized aged with varying degrees of cognitive impairment. They found that older persons did not respond readily to the GAT and that the instrument revealed only superficial aspects of the personality. Similarly, Fitzgerald, Pasewark, and Fleischer (1974) found that, from among a variety of themes, the GAT was more successful in eliciting responses from normal older individuals than the TAT only in the area of physical limitation. In another study, Pasewark, Fitzgerald, Dexter, and Cangemi (1976) found that the content of themes elicited by the TAT and GAT did not differ significantly between adolescent, middle aged and normal aged groups.

THE RORSCHACH TEST

The Rorschach test was originally developed as a tool for detecting psychopathology and in arriving at a diagnosis. Nevertheless, it is a measure of all of personality and has been used as a research and clinical tool in assessing personality in diverse populations. Klopfer and Taulbee (1976) have pointed out that the Rorschach test is one of the most widely used but also controversial projective techniques, as reflected in the diverse reviews given to it in the Mental Measurement Yearbook (e.g., Buros, 1972). On the one hand the Rorschach is seen as uniquely suited for providing a holistic assessment of perceptual, cognitive, and affective function and is routinely used by almost all clinical psychologists in their assessment batteries. On the other hand, its use as an adequate psychometric instrument has been widely criticized among researchers (Knutson, 1972). A legitimate perspective on the Rorschach may be that it provides a unique opportunity to obtain meaningful qualitative data about personality dispositions.

The Rorschach has been recommended for use with older persons because of the information it provides on inner strengths and weaknesses. Rather than stereotype the aged, this instrument allows clinicians to focus on individual psychodynamics. As early as 1954 Caldwell argued that as the scope of gerontology broadens to encompass personality research the Rorschach will serve as a valuable and major investigative technique.

Advantages in the use of the Rorschach with the aged have been presented by Klopfer (1974), Oberleder (1967), and Ames, Metraux, Rodell, and Walker (1973). These authors argue that the Rorschach is a relatively nonthreatening projective test. It

does not require a mode of performance which may penalize aged
persons whose coordination or speed of response has declined and
does not involve short term memory. At the same time the Rorschach is a highly ambiguous and unstructured test and its relevance may not be understood by older respondents. Klopfer (1974)
contends that aged persons may also have problems in seeing the
blot and that this may result in vague percepts. It should be
noted in this context that Eisdorfer (1960) studied the effects of
sensory impairment on Rorschach responses in the normal aged. He
found that hearing loss was related to poorer Rorschach scores.
Interestingly, visual impairment was not found to have a deleterious effect. A very important factor in considering the usefulness
of this test is the skill and training of the tester. This factor
is especially crucial with respect to the elderly who may not
readily respond to standard techniques of administration. Ames
(1974) recommends the Rorschach as a good "opener" in administering
a projective test battery to the aged. Based on extensive experience, she argues that the test is usually well received and is
seldom refused by the aged.

In contrast to the TAT which utilizes diverse and unstandar-
dized scoring systems, the Rorschach has a detailed and more generally employed scoring system. Consequently, norms for older
adults may be more readily reported and placed in context with
those of other groups. In considering and interpreting age norms,
however, one must be careful to recognize that deviations of older
persons from adult or middle-aged norms are not necessarily indications of maladjustment. Exner (1974) recently developed a comprehensive Rorschach scoring system to overcome the diversity in administration and scoring of the Rorschach, to improve comparability
of information gathered in clinical and research use, and to obtain
good reliabilities. It should be noted that there have not been
any reports of the use of this system with older populations as yet.

Rorschach Responses of the Aged

A number of early studies of institutionalized and community
aged have shown a relative paucity of m, k, K, and Fk responses
in these populations (Klopfer, 1974; Davidson and Kruglov, 1952).
Klopfer reports F% of the aged to be high while color is usually
absent. W responses are more frequently given than D responses.
It should be noted, however, that the above generalizations are not
based on longitudinal or even strictly quantified cross-sectional
comparisons. Older persons typically appear less productive on
the Rorschach (Davidson and Kruglov, 1952), giving a smaller number of responses per card. This observation is congruent with the
lower verbal output of the aged on the TAT and on other open-ended
instruments. In discussing this reduced productivity Klopfer

(1974) suggests that it may in fact be maturational, rather than regressive, and may be related to a general "cognitive load-shedding" by the aged. Based on a sample of the 97 women, Caldwell (1954) suggests that the meaning of the Rorschach responses are specific and that aspects of perceptual and intellectual functioning mediate responses of the aged to the Rorschach. She cautions against interpreting correlates of cognitive and perceptual changes as transformations of personality.

In contrast to the relative absence of longitudinal studies with the TAT a number of longitudinal studies do exist with respect to the Rorschach (Ames, 1965; Muller and LeDinh, 1976). Other studies have utilized the Rorschach to compare groups of elderly patients with varying degrees of mental impairments (Ames, et al., 1973). It should be noted that in a majority of these studies the Rorschach is not used as a projective test, but rather as a cognitive measure.

Developmental differences were referred to briefly even by Rorschach himself (1942). He suggested that the older person becomes more stereotyped and more constricted and loses the ability to utilize "inner resources." He also argued that responses of very old subjects show characteristics similar to those of demented patients. In another early study on the Rorschach, Klopfer (1946) concluded that the elderly show a slowing down in the intellectual sphere. Emotions are either restricted or labile and inner resources are considerably diminished. A 1947 study by Prados and Fried confirmed many of Klopfer's conclusions. It should be noted, however, that both were based on small samples. Another relatively earlier study by Davidson and Kruglov (1952) also pointed to age-related decrements.

The three studies noted above were reviewed and critiqued by Caldwell (1954) who argued that the inferences made by the authors about the decrements shown by the aged rest on a questionable assumption; that is, scoring categories developed for younger age groups are equally well suited for the aged. Thus Caldwell raises one of the most critical and thorny questions, one which personality research using projectives has still not answered. Among the major research studies of the aged which have been conducted by Ames (1960a, 1960b, 1965; Ames, et al., 1973) one was a longitudinal study of 61 older persons above age 70 (Ames, 1960a). Modal changes with increased age included decreases in total number of responses, in the variety of content categories, and in uses of color. Human and animal movement responses also decreased. Increases in animal responses were observed with increased age. All of these trends indicate a restriction of personality with aging. The result of a cross-sequential study (Ames, 1965) pointed to greater introversiveness of the oldest group at the retest time.

In reviewing Ames' work, Chown (1967) suggested that the decrease in productivity in responses by the aged may reflect a generalized cognitive decline; however, the greater introversiveness is likely to reflect a true personality change.

Diagnostic Use of the Rorschach

Although the Rorschach was originally devised to diagnose various types of psychopathology, its use with the aged has been largely restricted to assessment of cognitive impairment. In a recent review of Rorschach studies Ames et al. (1973) suggested that differential degrees of cognitive impairment, i.e., "developmental states," result in far greater differences in Rorschach scores than do age differences. Consequently Ames et al. (1973) have used score profiles on the Rorschach to categorize respondents into normal, presenile, and senile groups. They provide charts for categorization of responses into these categories. It should be noted, however, that no external validation of these categories presently exists. Nevertheless, data from Ames' (1974) calibration study utilizing a battery of psychological tests with institutionalized middle-aged and older respondents indicate that the Rorschach is the best single indicator of developmental status. When the elderly sample in this study was divided into subgroups of normal, intact presenile, medium and deteriorated presenile, statistically significant differences were also found in their responses to four other tests--the Bender-Gestalt, Gessel Incomplete Man Test, Monroe's Visual Tree, and the Color Tree Test.

Age differences in Rorschach responses of institutionalized aged reflect lower intellectual and emotional flexibility of the older individual and generally poor functioning of all respondents when compared to adult norms (Grosmann, Warshawsky, and Hertz, 1951). Studies by Kuhlen and Kerl (1951) and Chesrow, Wasika, and Reinitz (1949) involving the institutionalized aged also revealed intellectual and emotional deterioration among these groups. Muller and LeDinh (1976) conducted a longitudinal study of 30 schizophrenics and found relatively few age changes other than increases in animal and popular responses as a function of age.

Studies comparing institutionalized and community living aged generally suggest that community aged more closely approximate response pattern of the middle-aged (Klopfer, 1974). Community and institutional differences have also been observed by Ames et al. (1973).

One major problem in using the extensive literature on the Rorschach studies of older persons as normative data is a criticism voiced by both Eisdorfer (1963) and Light and Amick (1956); most

of these studies have not controlled for intelligence or institutionalization, thus confounding inferences about age-related differences or changes in Rorschach responses. Eisdorfer (1963)
attempted to overcome this problem by studying Rorschach responses
of forty-two well functioning community aged who were volunteers
in the Duke Geriatric Studies. Intelligence measures (WAIS) as
well as the Rorschach data were obtained from the respondents.
Norms obtained for this sample were compared with those of Klopfer
(1956), Ames, Learned, Metraux, and Walker (1954), Dörken and Kral
(1951), Chesrow et al. (1949), Light and Amick (1956), and Prados
and Fried (1947). Eisdorfer's study points to the importance of
intelligence in determining Rorschach responses of the aged.
Older subjects in the high IQ group (116+) did not differ significantly from younger subjects. Eisdorfer (1963) suggests that
Rorschach responses cited as pathognomic of aging may in fact
be characteristic of institutional status or the poorer cognitive
functioning of older persons.

THE HOLTZMAN INK BLOT TEST

A variation on the Rorschach test, the Holtzman Ink Blot Technique, has been proposed as useful in assessing reality testing
and anxiety as well as hostility and impulse control among the
elderly (Oberleder, 1967). The Holtzman technique may also be especially useful with older persons since responses are limited to
one per card. Although the Holtzman technique is not as widely
used in clinical practice as the Rorschach, its clinical potentials
are encouraging. Multivariate techniques show good validity of
the Holtzman in differentiating schizophrenics from normals, with
only 7% normals and 12% schizophrenics "misclassified."

In a recent study by Overall and Gorham (1972) the Holtzman
Ink Blot test was used to discriminate 300 Veterans Administration
domiciliary patients with and without organic brain syndrome.
(The results of this study are also discussed in Chapter 3.) Both
the Holtzman and the WAIS were able to differentiate middle-aged
and elderly patients and those with and without organicity. Age
patterns were different from patterns due to organicity. In comparison to the middle-aged, the older age group had higher scores
on location, rejection responses, and animal responses but lower
responses on color, shading, anatomy, and barrier responses. In
contrast the chronic brain syndrome group had higher scores for
integration and popular responses and lower scores of movement,
anxiety, and hostility than did the younger age group. The
authors suggest these results reflect cognitive rather than psychodynamic differences.

THE DRAW-A-PERSON TEST

The Draw-A-Person Test (DAP) offers a quick, nonverbal measure
of personality and cognitive functioning. It has both special advantages and special limitations in use with older persons. The
DAP has been popularized among clinicians by Machover (1952) and
is commonly included as part of a diagnostic test battery. Research utilizing the DAP has been less extensive than that dealing
with the Rorschach or TAT, but is still voluminous. The test was
developed as a personality measure expanding on the scope of the
Goodenough (1926) Draw-A-Man Test of intelligence. It is based on
the assumption that one's drawing of a person inevitably involves
a projection of one's body image, one's self-concept, and one's
manner of relating to the environment.

Use and Administration of the Draw-A-Person
Test with the Aged

The DAP has certain advantages in use with the aged over other
tests of cognition and personality. The directions are very simple
and verbal skills are not required in completing the test. The
subject is simply asked to draw a picture of a person. The test
requires very little time to complete compared with other standard
tests of personality or cognition. The test can be useful with regard to intellectual assessment, in formulating a diagnosis, and
in portraying psychodynamics. Nevertheless, Machover (1952) emphasizes that extreme caution must be exercised in the clinical use of
the DAP.   Thorough training in the use of the method as well as
extensive clinical experience are essentials for useful interpretations based on this instrument. Clinicians and researchers
have often been reluctant to use the DAP with the aged because it
was felt that impaired vision and psychomotor abilities would camo-
flage the psychodynamic picture. Nevertheless, even the blind
have been known to produce figure drawings which reflect an articulated body image (Machover, 1952). Clinicians may also learn
through the use of the DAP about sensory and motor problems not
ordinarily evident, and, conversely, perceptual-motor strengths
may be portrayed in patients who are suffering from obvious motor
handicaps.

It is valuable for clinicians to recognize potential signs of
common physical problems affecting the aged on the DAP. Arthritis
may result in poor lines and contours and difficulty in fine motor
coordination. Nevertheless, known arthritics are sometimes capable
of producing much better drawings than one would have anticipated.
Difficulties in vision may affect the connection of lines to one
another and the appropriate placement of facial features. Hand

tremors will be manifested in the line quality of the figure drawing. Examples are fragmented or bearded lines and outrightly un-
steady lines. The degree and kind of impairment incurred by these
clinical conditions are often portrayed in the figure drawings.
Dynamic explanations would be tempered and, in some cases, foregone
by the existence of the above conditions. Lakin (1960) suggests
that certain variables such as area, height, and centeredness of
the DAP are not readily influenced by sensory-motor impairments
in the aged.

The Draw-A-Person Test demands that the subject produce, on
his own accord, an articulated and integrated image of a person.
Among the institutionalized elderly, the inability or unwilling-
ness to do this test occurs more often than refusal on the Bender-
Gestalt test (Kahana, Dvorkin, Pruchno, and Zarker, 1977). Such
declinations to draw-the-person are, in their own right, revealing.
When reluctance does occur, the patient should be reassured and
told that drawing is a difficult task for many people. Patients
should be encouraged and coaxed to "give it a try" and not to
worry about the results. In extreme cases the examiner may
briefly show to the patient a variety of drawings done by other
people without providing any specific concrete model of a person
to the patient. Frequent encouragement and praise are very helpful
in facilitating DAP production among reluctant patients.

Scoring the DAP

The original and most widely used scoring system for the DAP
is the one developed by Goodenough (1926) for intellectual assessment. The scoring system gives points to the presence of body
parts, for the relationships of body proportions, and for the por-
trayal of a differentiated and integrated perception of the human
figure. Harris (1963) has restandardized and updated the original
Goodenough scoring system. Kahana et al. (1977) has proposed a
revision of the Harris-Goodenough scoring system for the human
figure drawing. The objective is to increase the reliability of
the scoring system and to make it more relevant for elderly persons. Drawings of 200 residents of homes for the aged are currently being correlated with other indices of cognitive functions.

A scoring system for assessing normal personality growth and
development through the Draw-A-Person Test has been developed by
Machover (1952). The system depicts stages of psychosexual development, self-concept, relationship to the environment, impulse
control, and coping with instincts and defenses.

Diagnostic and Clinical Use of the DAP

An early review of research findings based on the DAP was presented by Swensen (1957). His survey of research during a nine
year period did not provide support for most of the specific interpretations of figure drawings proposed by Machover. The reviewer
concluded, however, that negative research findings were consistently contrasted by extremely positive evaluations of experienced
clinicians regarding the usefulness of the DAP in personality
assessment. Swensen reconciled this discrepancy by arguing that
the figure drawings alone are not sufficiently reliable and do not
yield sufficient data to validly describe personality. However,
in combination with other indices as they are commonly used in
clinical practice, they yield valuable information. Cassel, Johnson, and Berns (1958) found a correlation of .33 among experienced
clinicians with regard to diagnoses based on the figure drawing
test. However, after discussion and agreement on scoring standards
among the clinicians, this correlation rose to .72. Schaeffer
(1964) states that certain figure drawings are always accurately
diagnosed while other drawings are always difficult to diagnose.
Harris (1963) states that the use of specific signs in arriving
at a diagnosis from a figure drawing has not been found useful,
whereas global judgments made by clinicians enabled them to distinguish between normal and selected clinical groups.

Jones and Rich (1957) related Goodenough's scores of the DAP
and height of the figure drawing to Wechsler-Bellevue scores among
40 male residents of homes for the aged (average age = 78.5).
They found a correlation of .65 between the Goodenough IQ and the
Wechsler-Bellevue IQ. These findings compare favorably to a
Goodenough-Binet correlation of .70 among children. Results
of this study support the use of the Goodenough scoring system as
a quick measure of intelligence among the aged. They also point
to a need for controlling for intelligence whenever the DAP is
used as a personality measure.

Strumpfer (1963) examined the relationship between poor figure drawings and psychological deficits as indicated by age and
chronicity among 81 hospitalized psychiatric patients. Age (range
20-53) was not significantly related to figure drawing performance.
Chronicity of diagnoses showed highly significant negative correlations with most figure drawing variables.

Institutional-Community Differences

Differences on the Draw-A-Person Test between institutionalized
aged and children in the community were examined by Lakin (1956).
Lakin's study considered personality dynamics, self-concept, and

body image in addition to cognitive functioning. Findings portrayed more constricted, "shorter," and less adequately centered
drawings among the institutionalized aged than among the community
children. The confounding of developmental factors, illness, and
community-institutional differences in this study present a major
problem in interpreting its results. In a later study Lakin (1960)
compared matched samples of community and institutionalized aged.
Consistent with the previous study, institutionalized aged showed
more restriction, shorter and less adequately centered drawings.
Lakin (1960) suggested that institutional environments may foster
dependency and exacerbate feelings of insignificance.

In a study of institutionalized VA patients Apfeldorf, Randolph,
and Whitman (1966) considered the relationship between figure drawing performance and contact maintained in the community. Centeredness of figures in the DAP was found to be significantly related to
furlough utilization in a group of 51 institutionalized VA patients.
This relationship held even when intelligence was controlled for.

Prediction of Illness and Death

A retrospective study of physical disorders using the DAP was
conducted by Harrower, Thomas, and Altman (1975). Drawings of 204
former medical students were examined 13 to 23 years later. At a
follow-up the group was divided into 102 experimentals (who developed hypertension, coronary heart disease, cancer, mental illness,
emotional disturbances, and suicide) and 102 controls who remained
in good health. Stance of the DAP figures was used to determine
the subjects' attitude toward the outside world. The authors concluded that the figure drawing had good predictive potential for
certain future disease states.

In Lieberman's (1965) study of psychological correlates of
impending death, decreasing complexity was observed over time in
figure drawings of the death-imminent group while the death-delayed
group showed no decline or some improvement in complexity of the
figure drawings.

Assessment of the Therapeutic Change
with the DAP

Wolk (1972) utilized the DAP as part of the House-Tree-Person
battery for assessing changes in therapy with older subjects. He
found significant pre- and post-therapy differences in his population. These included greater emphasis on eyes and nose and generally larger, although more diffuse, post-therapy drawings.
These improvements paralleled clinically observed improvements in

the following areas: depression, interpersonal relations, and
improvement subsequent to brain syndrome. A study by Modell (1951)
successfully used the DAP to investigate personality changes which
accompany psychotic regression and recovery subsequent to different
types of therapy. Butler and Lewis (1973) have pointed to the potential for therapeutic use of self-drawings of the aged.

Age and Cohort Differences

Figure drawings have been examined among various age groups
for differences in self-concept as well as cognition. Lakin (1956)
argues that changes in the DAP should occur throughout the life
span as body image develops and changes. Cognitive decline should
also be reflected in the DAP production. Saarni and Azara (1977)
compared the figure drawings of adolescents, young adults, and
middle-aged persons in terms of anxiety, aggressiveness, and other
personality characteristics. Adolescents portrayed significantly
more signs of anxiety than the two adults groups who, in turn, did
not differ significantly from one another on this variable. Tuckman, Lorge, and Zeman (1961) studied self-image as reflected in
figure drawings among 104 older adults. Findings based on both
community and institutionalized aged did not reveal any loss of
intactness among the aged. Gravitz (1969), in a study involving
328 men and 141 women, did not find differences in sex roles between middle-aged and the older subjects in his group. Another
study by Gravitz (1966) involving a larger sample and a broad
spectrum of ages (17-59) suggests a definite trend toward same
sex drawings with increasing age. These findings support Wolk's
(1972) contention that studies using the DAP reveal wide individual
differences which are superimposed on age and cohort effects.

Lorge, Tuckman, and Dunn (1954) compared figure drawings of
self by younger and older adults with regard to proportion, motor
coordination, and "depth." Drawings of older persons were considered to be less complete and less well-coordinated than those of
younger subjects and also revealed greater incidence of bizarre
responses. In a study by Gilbert and Hall (1962) age differences
over a wide age span (9-91) were observed in Goodenough intelligence ratings. Age differences were minimal under age 60 with a
more substantial decline observed in the over 60 age group. Drawings of older persons portrayed more transparencies, fragmentation,
and poor motor coordination.

Issues in Considering Research with the DAP

It is important to note in evaluating research on the DAP
that almost all the reported studies appeared in the 50's and early

60's. Work using this instrument has been relatively scarce in
recent years. Furthermore, almost without exception studies utilized special institutionalized populations. The differential meaning of certain "signs" of maladjustment on the Draw-A-Person Test
for people of different age groups has been discussed by Wolk
(1972) as well as by Saarni and Azara (1977). It also should be
noted that the DAP tasks are not equivalent across the studies
cited. In some studies subjects were asked to draw themselves
(Tuckman, Lorge, and Zeman, 1961; Butler and Lewis, 1973), whereas
in other studies they were requested to draw a person (Gilbert and
Hall, 1962).

OTHER PROJECTIVE TESTS

     In addition to the most commonly used projectives which have
been reviewed, a number of additional projectives should be briefly
mentioned as having potential value in personality assessment of
the aged.

The Sentence Completion Test

     The Sentence Completion Test (Holsopple and Miale, 1954) is a
theoretically and methodologically sound technique for assessing unconscious motivations. It has been generally useful in clinical
practice and research. It is also a particularly useful projective
test in that it affords respondents an opportunity to provide their
own meaning to statements while at the same time permitting the
use of an objective scoring system. It may be used without the
danger of social desirability biasing the responses. The structure
of the test is simple and it takes 5 to 10 minutes to complete,
thus making it a useful tool for the elderly. The test consists of
a series of sentence stems which the subject is asked to complete.
The response and the latency period are recorded. The test is
usually administered verbally to the subject, especially to older
respondents; however, paper and pencil formats have been developed.
The latter is not recommended for use with the elderly because of
possible problems with vision, writing, or in comprehension of the
task.

     Sentence completions have been used as indices of personality,
ego development, and adaptation among various younger age groups
(Pollack, 1966). Sentence completions also have been used with
older persons to assess personality, mental health, and adjustment
(Peck, 1959; Carp, 1967). Carp (1967) sought to determine age

differences in adjustment among the aged. She developed an objective scoring system for the Sentence Completion Test. This measure
was found to be suitable across ages, including the elderly. Good
test-retest reliability and good interrater agreement was reported.

Shimonaka and Marase (1975) in a Japanese study used the Sentence Completion Test to compare institutionalized older women with
those living with their families in the community in terms of
values and attitudes toward life. Respondents who lived with
their families portrayed more positive attitudes toward the family
and toward life and death. They also expressed more interest in
the future than those living in institutions. The sentence completion technique was also used by Kahana and Kahana (1973) with a
group of 120 institutionalized aged to assess preferences and
needs of the elderly. In a subsequent study (Kahana and Kiyak,
1976) a revised version of the scale was administered to 264 elderly individuals who had recently relocated to homes or special
housing for the aged. Five types of coping patterns were deline-
ated by the authors, using the Sentence Completion Test: (1)
instrumental; (2) intrapsychic; (3) affective; (4) escapist; (5)
resigned helplessness. The same scoring system was found to be
applicable to all sentence stems. The affective and instrumental
strategies was the predominant modes of adaptation across the four
sentence completion items used. Escape strategy and inability to
cope appeared less frequently. Rank order of appearance of these
strategies were similar at Time 1 and Time 2 with one difference;
intrapsychic responses appeared more frequently at Time 1 than at
Time 2. Evidence was found for inter-individual variation in
coping style as well as situation-specific patterns (Kahana and
Felton, 1976).

The Bender-Gestalt Test

The Bender-Gestalt Test (Bender, 1938) has been a widely used
projective and diagnostic test among younger age groups and with
pathological populations. Its primary use is in assessment of
perceptual motor development and perceptual organization, although
personality dynamics may also be inferred through its use (Hutt,
1969). The Bender-Gestalt has been reported to correlate with
various psychiatric problems, including organicity, schizophrenia,
mental retardation, and with the inability to cope with various
stresses (Koppitz, 1964; Tolor and Shulberg, 1963).

In the original validation sample using the Bender-Gestalt,
age was found to be significantly and positively related to Bender
scores. A scoring system developed by Hain (1964) has also re-

portedly differentiated patients with organic impairment from those
with functional psychiatric problems. Although Oberleder (1967)
recommends this system for use with the aged, it has not actually
been validated for aged populations. The Pascal-Suttell (1951)
scoring system has been commonly used with the aged in both clinical practice and research. Based on difficulties in obtaining
sufficiently high inter-rater reliability on Bender drawings of
older persons, modifications of the scoring system have been proposed by Kahana et al. (1977). The revised scoring system attempts
to provide consistent objective criteria (e.g., definable and
measurable deviations from angles, squares and other geometric patterns) rather than simple ratings of these errors. In addition,
errors were classified along a continuum rather than according to
a present-absent dichotomy.

In a study of clinically intact institutionalized aged Kral
and Wigdor (1963) found statistically significant correlations between Bender-Gestalt scores, hypertension, neurological findings,
and Rorschach responses.

Performance of small groups of senile and of healthy older
persons were compared on the Bender by Canter and Straumanis
(1969). The study pointed to a significantly greater number of
errors by seniles but also revealed error scores by one-fourth of
the normal aged. The Bender-Gestalt test was also used as a means
of assessing impairment in cognitive functioning with aged patients
after drug therapy with cyclandelate (Fine, Lewis, Villa-Londa, and
Blakemore, 1970). In Lieberman's (1965) classic study of psychological correlates of impending death among institutionalized aged
the Bender-Gestalt test appeared to be an accurate predictor of
impending death. When the death-imminent and the death-delayed
groups were compared over time, the death-imminent exhibited a
decline in the quality of the Bender reproduction and smaller
Bender-Gestalt drawings than the death-delayed group.

In a recent study by Kahana and Kiyak (1976) eight rating
scales were devised to measure the respondent's behavior while
performing the Bender-Gestalt and the Figure Drawing tasks. These
scales measured tension level, nature of affect, attitude toward
examiner, dependence on examiner, and distractability. Each pair
of behavior descriptions was followed by a four point scale.
Examiners were asked to describe the respondent's coping style with
respect to the respondent's criticisms of himself and of the test
situation or of other external circumstances (e.g., poor lighting,
visual problems). Scores obtained on behavioral observations on
the Bender-Gestalt and the Figure Drawing tests were found to be
correlated with information obtained using a self-report measure
(a coping scale) and content of responses on the Sentence Completion Test.

The Hand Test

This test is a projective measure requiring respondents to explain the meaning of hands which are presented in different positions on cards. It has been utilized by Panek, Sterns, and Wagner (1976) in an exploratory study with older persons. The authors recommend this instrument for use with the aged because it is meaningful to older persons, requires few perceptual motor skills, and may be administered in a short period (about 10 minutes). The Hand Test has previously been used successfully with the retarded (Wagner and Capotosto, 1966) and with the deaf (Levine and Wagner, 1974). Panek et al. (1976) report findings of depletion and constriction of personality among the elderly sample when compared to younger respondents. They are currently attempting to validate this test with larger numbers of older adults.

The Rosenzweig Picture Frustration Test

The Rosenzweig Picture Frustration test (Rosenzweig, 1935) is a semi-projective method aimed at revealing and measuring an individual's pattern of coping with a variety of potentially frustrating situations. It represents an attempt to operationalize clinical, psychoanalytically oriented concepts regarding frustration and consequent behavior. To the extent that the aged endure increasingly frustrating realities with fewer options for realistic resolutions, the Rosenzweig test should shed light on the dynamics of coping among the aged.

The test consists of 24 cartoon-like pictures, each of which portrays a mildly frustrating common situation. The situation is presented by the figure on the left. The person on the right is shown with a blank caption which the subject is to fill in with the first response that comes to mind. Rosenzweig points out that facial expressions and other individuating characteristics of the cartoon figures were deliberately omitted thereby encouraging an optimal projective situation. Responses are scored on two dimensions: direction of aggression and type of aggression.

This test has been used only minimally with the aged in the past, but both the task demands and the context of the test make it potentially useful with older persons. In administering the test to older persons, an interview format may be preferable to the questionnaire form.

Age norm and studies of age-related differences are available. During adolescence there is a rise in extraggression and ego-defense and a fall in intraggression and group conformity as the teenager goes through a period of identity seeking and rebellion

against authority (Rosenzweig and Braun, 1970; Rosenzweig, 1970).
As young adulthood begins there is a stabilization in all scoring
categories which is maintained through middle age (Rosenzweig,
1950). Adults 50 to 80 years old show a significant increase in
extraggression and ego-defense and a decrease in imaggression
and Group Conformity Rating. It must be noted, however, that
these results are drawn from a very small sample of older adult
subjects (Rosenzweig, 1952, in preparation).

CONCLUSIONS

When considering the use of projective tests with the aged as
an alternative to structured objective personality tests, a number
of strengths as well as weaknesses emerge. On the minus side, one
must reiterate the often noted problems in scoring, deriving norms,
and validating projective test responses. These problems appear to
be especially striking in research use of projectives. On the
other hand, projectives avoid many of the problems which confront
the aged in multiple choice and paper and pencil tests. Projectives are more readily understood and do not require complex verbal
discrimination or a good vocabulary level or understanding of cul-
ture-bound symbols. There is also less need to be test-wise as in
dealing with rating scales or in figuring out complex formats or
instructions. Since projectives are always administered in a
face-to-face interview situation, the trained examiner has a better
opportunity to understand special problems of the aged respondent
in taking the test than one would with a paper and pencil administration.

The relative independence of projective test performance from
educational attainment and the usefulness of even partially completed protocols also make the use of projective tests particularly well suited to aged populations. Since projective tests are
not heavily dependent on standardization and there is some flexibility in mode of administration, clinicians may readily choose
or adapt a projective test according to the demands of a particular
test situation. The examiner has the flexibility of varying the
length of the test (e.g., selecting certain TAT cards). The examiner may obtain varied observational and test response data on the
patient's performance and coping processes in a diverse set of
testing situations. While projective tests require that the subject provide his own structure to unstructured situations with
which he is presented, they also allow him to gracefully avoid
areas which he finds particularly embarrassing or painful. Thus
he may even ignore dealing with uncomfortable situations which are
portrayed on the stimulus card. These significant omissions are
nevertheless noted and utilized dynamically. Lastly, the phenome-

non of giving socially desirable responses, which the aged are particularly prone to do, is minimized with the use of projective
tests.

In view of the general reluctance among clinicians and researchers to use projective and psychomotor tests with the aged,
it is interesting to note that in a study of 200 institutionalized
and 50 community aged (Kahana and Kahana, 1976) the vast majority
of respondents were able to comply with the test demands to draw
Bender designs (94%) and Human Figure Drawing (76%). These findings are especially striking since a fair number of aged respondents (Mean age = 80) suffered from visual impairment and some
psychomotor problems. The vast majority of respondents (75%) appeared to be quite independent in performing the test. They con-
centrated well (87%) and were relatively noncritical or fault-
finding of the test situation (79%). They are also relatively relaxed (76%) and assumed a friendly and positive attitude toward
the test situation.

The important effects of psychosocial factors on test results
of the aging must, nevertheless, be considered (Oberleder, 1964).
Thus a total understanding of the aging person within the context
of his cultural climate as well as the test situation itself is
essential in interpreting projective as well as objective test
results. Typical psychological test results which point to older
persons as rigid, constricted, and/or senile may be in fact a
function of greater anxiety aroused by these tests in the older
individual. Traditional indices of bizarre or compulsive ideation
may in fact be viewed as adaptive responses to the social situation
of older persons.

Very little research has been done on the effects of social
and demographic background variables on the projective test performance of the aged. Yet, it is very important to note that
people with varying ethnic, cultural, and social class backgrounds
differ in their interpretation of desirable and stressful events
and in their behavioral as well as intrapsychic responses to such
events (Dohrenwend, 1966).

In considering the usefulness of various projective measures
as indices of personality among the aged, perhaps the greatest
problem lies in the lack of convincing data validating various test
scores as reflecting specific personality dimensions or dynamics.
Perhaps, due to the prevalence of cognitive impairment among aged
clinical populations or an unwarranted emphasis on cognitive rather
than personality features in dealing with the aged, many of the
existing measures have been validated against cognitive indices.
This author is hopeful that as the emotional life of older persons becomes a respectable concern of researchers and clinicians,

more work will be done on the use of projectives in terms of what they are best suited to do, i.e., revelation of personality dynamics. Only through more systematic use of the projectives by clinicians and researchers who have an appreciation for one another's skills will these instruments fulfill their potential value in facilitating personality assessment of the aged.

REFERENCES

Ames, L. B. Age changes in the Rorschach responses of a group of elderly individuals. Journal of Genetic Psychology, 1960, 97, 257–285. (a)

Ames, L. B. Age changes in the Rorschach responses of individual elderly subjects. Journal of Genetic Psychology, 1960, 97, 287–315. (b)

Ames, L. B. Changes in the experience balance scores on the Rorschach at different ages in the life span. Journal of Genetic Psychology, 1965, 106, 279–286.

Ames, L. B. Calibration of aging. Journal of Personality Assessment, 1974, 38, 507–529.

Ames, L. B., Learned, J., Metraux, R. W., and Walker, R. N. Rorschach responses in old age. New York: Harper Brothers, 1954.

Ames, L. B., Metraux, R. W., Rodell, J. L., and Walker, R. N. Rorschach responses in old age. New York: Brunner-Mazel, 1973.

Apfeldorf, M., Randolph, J., and Whitman, G. Figure drawing correlates of furlough utilization in an aged institutionalized population. Journal of Projective Techniques, 1966, 30, 467–470.

Bellak, L. The T.A.T., C.A.T., and S.A.T. in clinical use (3rd Edition). New York: Gruen & Stratton, 1975.

Bender, L. A visualmotor Gestalt test and its clinical use. New York: The American Orthopsychiatric Association, 1938.

Birren, J. E., Butler, R. N., Greenhouse, S. W., Sokoloff, L., and Yarrow, M. R. (Eds.). Human aging I: A biological and behavioral study (USPHS Publ. No. 986). Washington, D.C.: U.S. Government Printing Office, 1963.

Bradburn, N. Measures of psychological well-being. Working paper, National Opinion Research Center, Chicago, January, 1964.

Britton, J. H. Dimensions of adjustment of older adults. Journal of Gerontology, 1963, 18, 60–65.

Buros, O. K. (Ed.). The Seventh Mental Measurement Yearbook (Vol. 1). Highland Park, New York: Gryphon Press, 1972.

Butler, R. N., and Lewis, M. I. Aging and mental health: Positive psychosocial approaches. St. Louis: C. V. Mosby, 1973.

Caldwell, B. McD. The use of the Rorschach in personality research with the aged. Journal of Gerontology, 1954, 9, 316–323.

Canter, A., and Straumanis, J. J. Performance of senile and healthy aged persons on the BIP Bender Test. Perceptual and Motor Skills, 1969, 28, 695–698.

Carp, F. M. The application of an empirical scoring standard for a sentence completion test administered. Journal of Gerontology, 1967, 22, 301–307.

Cassel, R., Johnson, A. P., and Burns, W. H. Examiner, ego-defense, and the H-T-P Test. Journal of Clinical Psychology, 1958, 14, 157–160.

Chesrow, E. J., Wasika, P. H., and Reinitz, A. H. A psychometric evaluation of aged white males. Geriatrics, 1949, 4, 169–177.

Chown, S. Personality and aging. In K. W. Schaie (Ed.), Theory and methods of research on aging. Morgantown, W.V.: West Virginia University Library, 1967.

Davidson, H. H., and Kroglov, L. Personality characteristics of the institutionalized aged. Journal of Consulting Psychology, 1952, 16, 5–12.

Dohrenwend, B. Social status and psychological disorder: An issue of substance and an issue of method. American Sociological Review, 1966, 31, 14–34.

Dörken, H., and Kral, A. V. Psychological investigation of senile dementia. Geriatrics, 1951, 6, 151–163.

Eisdorfer, C. Developmental level and sensory impairment in the aged. Journal of Projective Techniques, 1960, 24, 129–132.

Eisdorfer, C. Rorschach performance and intellectual functioning in the aged. Journal of Gerontology, 1963, 18, 358–363.

Exner, J. E. Jr. The Rorschach: A comprehensive system. New York: Wiley Interscience, 1974.

Fine, E. W., Lewis, D., Villa-Londa, I., and Blakemore, C. B. The effect of cyclandelate on mental function in patients with arteriosclerotic brain disease. British Journal of Psychiatry, 1970, 117, 157–161.

Fink, H. The relationship of time perspective to age, institutionalization and activity. Journal of Gerontology, 1957, 12, 414–417.

Fitzgerald, B., Pasewark, R., and Fleischer, S. Responses of an aged population on the Gerontological and Thematic Apperception Tests. Journal of Personality Assessment, 1974, 38, 234–275.

Gaitz, C. M. Mental disorders: Diagnosis and treatment. In E. Busse (Ed.), Theory and therapeutics of aging. New York: Medcom Press, 1973.

Gaitz, C. M., and Baer, P. E. Diagnostic assessment of the elderly: A multifunctional model. The Gerontologist, 1970, 10, 47–52.

Gilbert, J., and Hall, M. R. Changes with age in human figure drawing. Journal of Gerontology, 1962, 17, 397–404.

Goodenough, F. L. Measurement of intelligence by drawings.

Yonkers, New York: World Book Company, 1926.

Granick, S., and Friedman, A. S. The effects of education on the decline of psychometric test performance with age. Journal of Gerontology, 1967, 22, 191-195.

Gravitz, M. A. Normal adult differentiation patterns on the Figure Drawing Test. Journal of Projective Techniques, 1966, 30, 471-473.

Gravitz, M. A. Marital status and figure drawing choice in normal older Americans. Journal of Social Psychology, 1969, 77, 143-144.

Grossman, C., Warshawsky, F., and Hertz, M. Rorschach studies on personality characteristics of a group of institutionalized old people. Journal of Gerontology, 1951, 6 (Suppl. 3), 97. (Abstract)

Guilford, J. P. (Ed.). Fields of psychology, basic and applied. New York: Van Nostrand, 1950.

Gutmann, D. L. The country of old men: Cross cultured studies in the psychology of later life. In Occasional papers in gerontology. University of Michigan and Wayne State University, April, 1969.

Gutmann, D. L. Alternatives to disengagement: The old Maya and the Highland Druze. In J. F. Gubrium (Ed.), Time, roles, and self in old age. New York: Human Science Press, 1976.

Hacker, S., Gaitz, C., and Hacker, B. A humanistic view of measuring mental health. Journal of Humanistic Psychology, 1972, 12, 94-105.

Hacker, S., Gaitz, C., and Hacker, B. Measuring mental health and illness: Analysis of empirical relationships between measurements of concepts. Unpublished manuscript, Texas Research Institute of Mental Sciences, Houston, 1970.

Hain, J. D. The Bender Gestalt Test. A scoring method for identifying brain damage. Journal of Consulting Psychology, 1964, 28, 34-40.

Harris, D. B. Children's drawings as measures of intellectual maturity. New York: Harcourt, Brace & World, Inc., 1963.

Harrower, M., Thomas, C. B., and Altman, A. Human figure drawings in a prospective study of six disorders: Hypertension, coronary heart disease, malignant tumor, suicide, mental illness, and emotional disturbance. Journal of Nervous and Mental Disease, 1975, 161, 191-199.

Hartman, A. A. A basic TAT set. Journal of Projective Technology and Personality Assessment, 1970, 34, 391-396.

Henry, W. E. The analysis of fantasy: The Thematic Apperception Technique in the study of personality. New York: John Wiley & Sons, Inc., 1956.

Hochschild, A. R. Disengagement theory: A critique and proposal. American Sociological Review, 1975, 40, 553-569.

Holsopple, J., and Miale, F. Sentence completion--a projective method for the study of personality. Springfield, Il:

Charles C Thomas, 1954.

Howell, S. Assessing the function of the aging adult. The Gerontologist, 1968, 8, 60-62.

Hutt, M. The Hutt adaptation of the Bender Gestalt Test (2nd ed.).
New York: Gruen & Stratton, 1969.

Jones, A. W., and Rich, T. A. The Goodenough Draw-A-Man test as a
measure of intelligence in aged adults. Journal of Consulting
Psychology, 1957, 21, 235-238.

Kahana, B., Dvorkin, L., Pruchno, R., and Zarker, T. Adaptation of
clinical assessment techniques for use with the aged. Paper
presented at the Gerontological Society Annual Meeting, San
Francisco, 1977.

Kahana, B., and Kahana, E. Changes in mental status of elderly
patients in aged-integrated and age-segregated hospital
milieus. Journal of Abnormal Psychology, 1970, 75, 177-181.

Kahana, B., and Kiyak, A. The use of projective tests as an aid
to assessing coping behavior and coping styles among the aged.
Paper presented at the 29th Annual Meeting of the Gerontological Society, New York, 1976.

Kahana, E., and Kahana, B. Effects of age segregation on affective expression of elderly psychiatric patients. American
Journal of Orthopsychiatry, 1968, 38, 1968.

Kahana, B., and Kahana, E. Strategies of adaptation to institutional environments. Symposium given at the Annual Meeting
of the Gerontological Society, Miami, 1973.

Kahana, E., and Kahana, B. Strategies of coping in institutional
environments. Progress Report submitted to National Institute of Mental Health, 1976.

Kahana, E., and Felton, B. J. Continuity and change in coping
strategies in a longitudinal analysis. Paper presented at
the 29th Annual Meeting of the Gerontological Society, New
York, 1976.

Klopfer, W. G. Personality patterns of old age. Rorschach
Research Exchange, 1946, 10, 145-166.

Klopfer, W. G. The application of the Rorschach Technique to
geriatrics. In W. G. Klopfer (Ed.), Developments in the
Rorschach technique, Fields of application (Vol. 2). Yonkers, New York: World Book Co., 1956.

Klopfer, W. G. The Rorschach and old age. Journal of Personality
Assessment, 1974, 38, 420-422.

Klopfer, W. G., and Taulbee, E. S. Projective tests. Annual Review of Psychology, 1976, 27, 543-567.

Knutson, J. F. Rorschach. In O. K. Buros (Ed.), Mental Measurement Yearbook (Vol. 1). Highland Park, New York: Gryphan,
1972.

Koppitz, E. M. The Bender Gestalt test for young children. New
York: Gruen & Stratton, 1964.

Kral, V. A., and Wigdor, B. T. Clinical and psychological observations in a group of well-preserved aged people. Medical

Services Journal of Canada, 1963, 19, 1-11.

Kuhlen, R. G., and Kerl, C. The Rórschach performance of 100 elderly males. Journal of Gerontology, 1951, 6 (Suppl. to No. 3), 115. (Abstract)

Lakin, M. Certain formal characteristics of human figure drawings by institutionalized aged and by normal children. Journal of Consulting Psychology, 1956, 20, 471-474.

Lakin, M. Formal characteristics of human figure drawings by institutionalized and non-institutionalized aged. Journal of Gerontology, 1960, 15, 76-78.

Lehmann, H. E., Ban, T. A., and Kral, V. A. Psychological test practice effect in geriatric patients. Geriatrics, 1968, 160-163.

Levine, E. S., and Wagner, E. E. Personality patterns of deaf persons: An interpretation based on research with the Hand Test. Perceptual and Motor Skills, 1974, 39, 1167-1236 (Mongr. Suppl. 1-V39).

Lewin, K. Field theory in social science. New York: Harper, 1951.

Lieberman, M. A. Psychological correlates of impending death. Some preliminary observations. Journal of Gerontology, 1965, 20, 182-190.

Lieberman, M. A., and Caplan, A. S. Distance from death as a variable in the study of aging. Developmental Psychology, 1970, 2, 71-84.

Light, B. A., and Amick, J. H. Rorschach responses of normal aged. Journal of Projective Techniques, 1956, 20, 185-195.

Lorge, I., Tuckman, J., and Dunn, M. B. Human figure drawing by younger and older adults. American Psychologist, 1954, 9, 420-421. (Abstract)

Machover, K. Personality projection in the drawing of the human figure. Springfield, Ill.: Charles C Thomas, 1952.

Modell, A. H. Changes in human figure drawings by patients who recover from regressed states. American Journal of Orthopsychiatry, 1951, 21, 584-596.

Muller, C., and LeDinh, T. Aging of schizophrenic patients as seen through the Rorschach Test. Acta Psychiatrica Scandinavica, 1976, 53 (Issue 3), 161-167.

Neugarten, B. L., and Gutmann, D. L. Age-sex roles and personality in middle age: A thematic apperception study. In B. L. Neugarten (Ed.), Middle age and aging. Chicago: Atherton Press, 1968.

Neugarten, B. L., Havighurst, R. J., and Tobin, S. S. The measurement of life satisfaction. Journal of Gerontology, 1961, 16, 134-143.

Oberleder, M. Effects of psychosocial factors on test results of the aging. Psychological Reports, 1964, 14, 383-387.

Oberleder, M. Adapting current psychological techniques for use in testing the aged. Gerontologist, 1967, 7, 188-191.

Overall, J. E., and Gorham, D. R.   Organicity versus old age in objective and projective test performance. Journal of Consulting and Clinical Psychology, 1972, 39, 98-105.

Panek, P. E., Sterns, H. L., and Wagner, E. E.   An exploratory investigation of the personality correlates of aging using the hand test. Perceptual and Motor Skills, 1976, 43, 331-336.

Pascal, G. R., and Suttell, B. J.   The Bender-Gestalt Test: Quantification and validity for adults. New York: Gruen & Stratton, 1951.

Pasewark, R., Fitzgerald, B., Dexter, V., and Cangemi, A.   Responses of adolescent, middle aged and aged females on the Gerontological & Thematic Apperception Tests. Journal of Personality Assessment, 1976, 40, 588-591.

Peck, R. F.   Personality factors in adjustment to aging. Geriatrics, 1960, 15, 124-130.

Pollack, D.   Coping and avoidance in inebriated alcoholics and normals. Journal of Abnormal Psychology, 1966, 71, 417-419.

Postema, L. Y.   Reminiscing, time orientation and self concept in aged men. Unpublished doctoral dissertation, Michigan State University, 1970.

Prados, M., and Fried, E. G.   Personality structure in the older age groups. Journal of Clinical Psychology, 1947, 17, 302-304.

Rorschach, H.   Psychodiagnostics. New York: Grune and Stratton, 1942.

Rosen, J. L., and Neugarten, B. L.   Ego functions in the middle and later years: A thematic apperception study of normal adults. Journal of Gerontology, 1960, 15, 62-67.

Rosenzweig, S.   A test for types of reactions to frustration. American Journal of Orthopsychiatry, 1935, 5, 395-403.

Rosenzweig, S.   Revised norms for the adult form of the Rosenzweig Picture Frustration Study. Journal of Personality, 1950, 18, 344-346.

Rosenzweig, S.   Laboratory reports on miscellaneous Picture Frustration Study projects. Unpublished manuscript, 1952.

Rosenzweig, S.   Sex differences in reaction to frustration among adolescents. In J. Zubin and A. Freedman (Eds.), Psychopathology of adolescence. New York: Gruen & Stratton, 1970.

Rosenzweig, S., and Braun, S. H.   Adolescent sex differences in reactions to frustration as explored by the Rosenzweig P-F Study. Journal of Genetic Psychology, 1970, 116, 53-61.

Rosenzweig, S.   Aggressive behavior and the Picture Frustration study. Book manuscript in preparation.

Saarni, C., and Azara, V.   Developmental analysis of human figure drawings in adolescence, young adulthood, and middle age. Journal of Personality Assessment, 1977, 41, 1, 31-38.

Schaeffer, R. W.   Clinical psychologists' ability to use the Draw-A-Person Test as an indicator of personality adjustment.

Journal of Consulting Psychology, 1964, 28, 383. (Abstract)

Schaie, K. W. Age changes and age differences. The Gerontologist, 1967, 7, 128-132.

Schaie, K. W., and Schaie, J. P. Clinical assessment and aging. In J. E. Birren and K. W. Schaie (Eds.), The handbook of the psychology of aging. New York: Van Nostrand Reinhold Co., 1977.

Shrut, S. Attitudes toward old age and death. Mental Hygiene, 1958, 42, 259-266.

Shimonaka, Y., and Marase, T. Japanese Journal of Educational Psychology, 1975, 23, 104-113.

Strumpfer, D. J. The relations of DAP Test variables to age and chronicity in psychotic groups. Journal of Clinical Psychology, 1963, 19, 208-211.

Swensen, C. H. Empirical evaluations of human figure drawings. Psychological Bulletin, 1957, 54, 431-466.

Tolor, A., and Shulberg, H. An evaluation of the Bender-Gestalt Test. Springfield, Ill.: Charles C Thomas, 1963.

Traxler, A., Swiener, R., and Rogers, B. Use of Gerontological Apperception Test (GAT) with community-dwelling and institutional aged. The Gerontologist, 1974, 14, 52. (Abstract)

Tuckman, J., Lorge, I., and Zeman, F. D. The self image in aging. Journal of Genetic Psychology, 1961, 99, 317-321.

Veroff, J., Atkinson, J. W., Feld, S. C., and Gurin, G. The use of thematic apperception to assess motivation in a nationwide interview study. Psychological Monographs, 1960, 74, (12, Whole No. 499).

Wagner, E. E., and Capotosto, M. Discrimination of good and poor retarded workers with the Hand Test. American Journal of Mental Deficiency, 1966, 1, 126-128.

Wolk, R. L. Refined projective techniques with the aged. In D. P. Kent, R. Kastenbaum, and S. Sherwood (Eds.), Research planning and action for the elderly: The power and potential of social science. New York: Behavioral Publications, 1972.

Wolk, R. L., and Wolk, R. B. The Gerontological Apperception Test. New York: Behavioral Publications, 1970.

PSYCHOSOMATIC ISSUES IN ASSESSMENT

W. Doyle Gentry

Duke University Medical Center

The last decade and a half has witnessed the emergence of the field of clinical gerontology and a rapidly growing interest in the physical and mental health problems of the aged. In a few short years, lay and professional groups have been organized, political interest has become focused, governmental task forces have been commissioned, professional journals have appeared, books have been written, subspecialties in several fields of medicine and social science have been developed--all centering around a basic concern with the nature and scope of health care for the elderly.

Perhaps the most recent interest in health problems of the aged has been expressed by clinical psychologists (Gentry, 1977). Lagging behind the professions of psychiatry, nursing, sociology, and social work, clinical psychology has only just begun to consider the possibilities and opportunities for applying psychodiagnostic and therapeutic techniques to the health needs of the elderly who reside in nursing homes and mental health institutions, as well as those living in the community. To date, this interest has for the most part been restricted to interaction with older persons in community mental health programs, in using behavior modification techniques to treat mental disorders, in evaluating the effectiveness of mental health programs for the aged, and in utilizing standardized psychological tests to identify intellectual deficit, personality disorders, and organicity.

The fact that clinical psychologists have thus far not demonstrated any interest in the relationship of psychological factors (stress, personality traits, coping mechanisms) to physical disorders, i.e., _psychosomatic_ disorders, in the aged is not surprising. This same absence of interest in such relationships charac-

terizes both geriatric medicine and geropsychiatry, disciplines
which have focused exclusively either on the mental or physical
health problems of older persons. Much is currently known about
the increased physical morbidity associated with aging and the increased incidence of disorders such as arteriosclerosis, cardiovascular disease, impairment of vision and hearing, hypertension,
pulmonary disease, and painful arthritis (Dovenmuehle, Busse, and
Newman, 1970). Likewise, a great deal is known about the extent
of psychiatric disorders in the elderly (Whanger and Busse, 1976),
including disorders such as depression, dementia, manic-depressive
disorders, hypochondriasis, and organic brain syndromes. However,
almost no reference is made in the relevant literature to the rela-
tionship(s) between psychological and physical factors in assessing
the health care needs of the aged. To illustrate this latter point,
one should consider that: (1) Butler and Lewis (1973) devote only
a half-page discussion to "psychophysiological or psychosomatic
disorders" in their excellent book on Aging and Mental Health; (2)
Nowlin and Busse (1977) in discussing the psychosomatic problems in
the older person are limited to a description of case studies illus-
trating hypochondriasis and depression, primarily in patients evi-
dencing chronic pain symptomatology; and (3) a review of Index
Medicus references to journal articles in the field of gerontology
for the past ten years yielded only a handful of citations under
the heading of psychosomatic or psychophysiologic disorders, most
of which described studies dealing with cancer and other terminal
illnesses, organic brain syndromes, and psychological reactions of
aged patients to various medical stresses.

     The aim of the present chapter is to extend the discussion of
how psychological and physical factors are associated in disorders
affecting the elderly, including both "psychosomatic" relationships
where psychological factors cause or contribute to physical symptoms
and also "somatopsychic" relationships where physical problems of
the aged person lead to certain psychological reactions, and to
address issues concerning the assessment of these relationships in
older persons. Also considered will be whether or not psychosomatic disorders are more or less prevalent among the elderly, as
compared to persons in their early or middle years, and how psychosomatic disorders differ from disorders such as hypochondriasis
and hysteria (conversion), the latter frequently discussed by mental
health professionals interested in gerontology.

Psychosomatic Disorders In The Elderly

     Butler and Lewis (1973) indicate that some of the more common
psychophysiologic disorders found in older persons include: psychogenic rheumatism, hyperventilation syndromes, irritable colon,
pruritus (itching) ani and vulvae, preoccupation with bowel dysfunction, and nocturia (excessive urination at night). Less common,

they suggest, are disorders such as ulcers which are stress-pro-
duced.

Weiner (1977) notes that many psychosomatic disorders begin
in a setting of grief and bereavement, a psychological state
familiar to many older people. The individual is usually grieving
over a loss, either real or threatened, including such things as
separation from children, death of a spouse or friend, loss of
function due to traumatic illness, loss of a job (retirement), and
loss of possessions (e.g., being moved out of one's home of thirty
years into a nursing home). The individual feels helpless to cope
with the situation and is enveloped by a sense of hopelessness and
depression. Weiner indicates that grief may result in increased
morbidity for diseases such as cancer, diabetes, tuberculosis, and
congestive heart failure, as well as psychosomatic disorders such
as essential hypertension, peptic ulcers, and rheumatoid arthritis.

As Lindemann (1977) suggests, morbid grief reactions may often
be delayed and distorted in character. That is, a person may manifest sudden symptoms of a somatic illness many years after the
death (loss) of a parent or loved one. The onset of illness, e.g.,
symptoms of coronary heart disease, may in fact often appear on
the anniversary of the loss of the individual for whom they are
still grieving, hence the term anniversary reaction. The illness
symptoms will frequently mimic those which characterized the death
of the person for whom they are grieving, e.g., the father who
died prematurely from sudden heart attack at age 52, and lack any
organic basis. This condition is not uncommon in patients seeking
medical treatment in their later years.

As Lindemann also indicates, sometimes the psychophysiologic
disorder manifested by the patient appears in anticipation of
some loss, i.e., results from anticipatory grief. Such reactions
again are not uncommon in the elderly, e.g., an older parent who
is losing an only child to marriage or relocation for job purposes (separation) or an elderly individual married to someone
with terminal cancer. While the anticipatory grief reaction may
have some "survival value" in that it prepares and to some extent
protects the older person from the reality of the inevitable loss,
it nevertheless frequently results in unpleasant and potentially
dangerous somatic illness such as ulcerative colitis, high blood
pressure, asthma, and rheumatoid arthritis. Lindemann notes, for
example, that in a study of patients with ulcerative colitis,
80% developed their disease in close relationship to the loss of
some important person. Butler and Lewis (1973) point out that the
physical concomitants of grief include a feeling of emptiness in
the pit of the stomach, weak knees, feelings of suffocation,
shortness of breath, and a tendency for deep sighing. They also
note that, among older persons undergoing a grief reaction, it is
not uncommon to find evidence of insomnia, digestive disturbances,
and anorexia.

A typical clinical illustration of anticipatory grief reaction in an elderly person is as follows:

> This 52-year-old female was seen for psychological evaluation to evaluate possible organicity vs. depressive symptomatology. At the time of interview, she was friendly, cooperative and highly motivated towards "finding out what is wrong" with her. Her thought content centered primarily around her various somatic problems and her inability to remember things anymore. She had a prior history of migraine headaches and recently had been experiencing "spells" in which she experienced blurred vision, dizziness, and severe headaches. She also complained of being generally upset, of being extremely weak and of losing interest in previously enjoyable past-times such as reading and sewing. She stated that her head often felt "heavy," as if it were a "dead weight." She could not concentrate, forgot to do things, and could not remember either routine things or specific events from her remote past. She often felt as if "things are closing in on me" and became extremely anxious about such feelings.
>
> She worried a great deal about her husband and his health. He was from a family in which all members suffered from cardiovascular disease. Most family members were eventually stricken with sudden heart attacks or strokes (CVA). The patient stated that she always had lived with the fear that her husband, who had hypertension, also would be stricken in the same sudden manner. While relating this fear about her husband's welfare, she became quite upset and cried. Her husband refused to discuss his health status with the patient, so she was forced to "keep it bottled up inside" of herself. She described herself as left with the feeling that "there is nothing I can do."
>
> The patient always had had a feeling that "she was strong enough to carry any load," but now just didn't know if she could cope with this situation. She did not want to let herself get "down" because her husband "needs" her. At one point in the interview, she suddenly asked the interviewer, "Am I going to have a stroke?"
>
> During the testing procedures the patient did not appear to be anxious and participated seemingly with a high degree of interest. She did not appear to be fatigued at any time despite the rather lengthy testing session and persisted on all tasks given to her. She responded

quickly to all questions and did not give up easily.

Her general emotional level throughout the interview and
testing was best described as depressed. She would become extremely distraught when discussing her problems
and appeared somewhat relieved after "letting it all
out." The interview had definite cathartic value for
this patient.

Analysis of Test Data. This patient was functioning
within the high "Average" range of intellectual ability.
She attained a full scale WAIS I. Q. of 109, with her
verbal skills being slightly better (Verbal I.Q. = 113)
than her performance skills. Of some import is the fact
that she obtained extra credit on several test items for
using minimal time in successfully completing the task.

On the Benton Visual Retention Test, her performance was
extremely good and showed no signs of any decrement
associated with organic brain damage. The same was true
of her performance on the Trail Making Test from the
Halstead-Reitan neuropsychological battery.

On the Wechsler Memory Scale (Form I), she attained a
memory quotient of 128, which falls within the high
"Superior" range and is in keeping with her verbal I.Q.
She appeared to function adequately in all areas of
this test with no noticeable deficiencies.

Thus, on all tests dealing with intellectual or memory
function, this patient exhibited no signs of decrement
of the type usually associated with organic impairment.
On the contrary, she manifested above average and
superior skills in these areas. This was somewhat
surprising in view of her symptomatology involving memory loss and an inability to concentrate.

With regard to personality structure, the picture was
less encouraging. Her response to the MMPI indicated
that she was utilizing psychoneurotic defense mechanisms
in an effort to handle her feelings of frustration and
hostility. Her primary coping mechanism was repression,
a defense she used to deny her "fears of losing her husband" and her anger towards him for not appreciating
her concern, i.e., discussing the situation with her.
This defense was not effective, however, and had resulted in a state of depression, obsessive ruminations
about her somatic symptoms, emotional lability and a
self-critical attitude. She incorporated her husband's
symptomatology through the process of identification and

was convinced that she was going to have a stroke. She
demonstrated all the hysterical symptoms associated
with this type of cardiovascular disorder: numbness
over various regions of the body, especially in the left
arm; tingling in the legs; cold claminess of the left
extremities. At the time of evaluation she was at a
point of low "defensiveness" with insufficient ego-
strength to deal with her problems. She felt that
things were "closing in" on her and consequently was
very anxious, depressed and unsure about her future. The
degree of anxiety was such that it, at times, clouded her
thought processes and severely disrupted her otherwise
superior intellectual and memory faculties. Such difficulties in turn reinforced her feelings of insecurity and
enhanced her depression and self-castigation.

The patient appeared to be a good candidate for psychotherapy. She needed to be able to openly express her
hostility and frustration and be reinforced for such behavior. If she were able to talk her fears out about her
husband, perhaps they would not build up to the proportions of their present state. She was seen as requiring
a lot of supportive care while attempting to deal with
such feelings, but the prognosis for remission of symptoms was thought to be good.

Nowlin and Busse (1977) further emphasize the interaction between the older person's loss of physical reserve (e.g., decreased
cardiac output, kidney function, and cerebral blood flow) and the
attrition older persons experience in losing family members and
friends through death, retirement from gainful employment, and
diminished physical vigor, which directly contributes to increased
psychosomatic illness. They note that psychophysiologic symptoms
in the elderly are often related to concurrent depression, e.g.,
in patients with chronic musculoskeletal pain in the neck and lower
back region. Often they regard the somatic illness presented by
the older patient as a depressive equivalent and treat it with antidepressant medication; as the depression lifts, the somatic symptoms lessen and/or disappear altogether.

Distribution of Psychosomatic Disorders As A Function of Age

One can certainly get the impression that psychosomatic disorders are extremely prevalent in older persons being seen by medical
and psychiatric specialists. This is evident in the observation
that many of these disorders are common among older persons and the
fact that such disorders often result from loss and are thus associated with grief and depression, conditions all too frequently seen
in the elderly. However, there is little data available to support

this impression of increasing incidence of psychosomatic illness as a function of age.

One study by Engelsmann, Murphy, Prince, Leduc, and Demers (1972), in fact, suggests that this is not the case. These investigators surveyed a total of 875 individuals with the Langner symptom check-list and analyzed the responses by sex, age, income, residence (rural vs. urban) and ethnicity. Interestingly, while all of the other variables seem to relate to differences in various clusters of symptoms (including a psychosomatic cluster), the variable of age did not. The psychosomatic cluster included the following somatic complaints: breathing difficulties, feeling hot, heart pounding, weakness, fainting spells, acid stomach, cold sweats, and headaches. They concluded that within the age range of 18 to 60 years, the variable of age can be ignored in understanding the distribution of such illnesses among a Canadian population.

In a recent study at Duke University Medical Center[1] a comparison was made between two different age groups of mental health clients responding to a comprehensive computer interview (Angle, Ellinwood, Hay, Johnson, and Hay, 1977). This interview inquired about problems in over 250 areas of potential difficulty including several dealing with health concerns and physical disorders which may be considered under the general rubric of psychosomatic disorders. Clients were divided into those above (n = 51) and below (n = 389) fifty-five years of age; dividing the sample at a higher age provided too few subjects in the older group for statistical comparison. Comparison of the patients' responses to the various problem categories by means of t-tests revealed the following significant differences:

1. Older clients reported more frequent contacts with physicians and hospitals both during the previous thirty days and over the previous year than did younger clients.
2. Older clients reported more sleep problems involving: waking up frequently, difficulty in getting back to sleep, restlessness, awakening before becoming rested, the need for and use of sleep medications, and having physical problems which interfered with sleep (e.g., bladder problems).
3. Older clients evidenced more health worries than did younger clients and had experienced health worries over a longer period of time.
4. Older clients reported a greater degree of social isolation, i.e., spending time alone and away from friends, because of poor health and physical disability.

[1]These data were provided by Dr. Hugh V. Angle and analyzed by Ms. Judy Carroll through a NIDA Grant No. 1H81DA01665.

5. Older clients indicated that their interpersonal relationships with family members were more characterized by excessive somatic complaints than was true of younger clients.
6. Older clients reported being bothered more by pain during
   the last 90 days; they described their pain as more frequent, more intense, and as lasting for longer periods of
   time than did younger clients. There was also a greater
   indication that the pain symptoms of older clients was
   operant in nature, i.e., was greatly reinforced by environmental relationships. For example, older clients more
   often stated that their pain began only when they engaged
   in some type of physical activity and/or they frequently
   had others to help or assist them when their pain occurred.
   They also reported a greater restriction of activities because of the pain, especially interruptions in sleep.
7. Older clients reported more episodes of musculoskeletal
   pain, bladder problems (passing blood in the urine, excessive urination, burning pain during urination, loss of
   bladder control), poorer physical health status, and a more
   significant illness history. The latter category included
   reference to disorders such as: diabetes, cancer, tuberculosis, heart trouble, rheumatoid arthritis, stroke, and
   other chronic illnesses.
8. Finally, in ranking categories which the clients considered moderately to severely problematic, older clients
   ranked "medical problems" and "sleep disorders" much
   higher than did younger clients.

In summary, there did appear to be both a greater preoccupation with physical health concerns among the older clients and also
more evidence of actual somatic symptomatology. From a psychosomatic viewpoint, this is significant since it must be remembered
that these clients were psychiatric patients referred from a mental
health facility and not simply a group of normal aged persons.
Many were manifesting depressive reactions and other mental disorders which required hospitalization, yet their primary concerns
and the actual manifestations of the psychological difficulties
were in somatic terms. Physical health problems resulted in a
greater degree of social isolation and characterized their relationships with family and friends, i.e., excessive somatic complaints
and assistance from others because of pain/disability. These data
clearly support the earlier observations of Butler and Lewis (1973),
Weiner (1977), and Nowlin and Busse (1977) regarding the types of
psychosomatic syndromes frequently seen in the elderly, e.g.,
rheumatism, bladder dysfunction, nocturia, and so forth.

The conflicting results of the above two studies suggest the
need for further research in this area before any definitive conclusions can be drawn as to whether or not psychosomatic illness
is more prevalent in the elderly.

Somatopsychic Disorders in the Elderly

The psychological consequences of physical illness in older persons has received considerable attention to date. Verwoerdt (1976), for example, has noted that the type and intensity of soma-topsychic reactions in the elderly depends on: (1) the stresses of the physical illness per se; (2) the characteristics of the "host" person; and (3) situational factors surrounding the illness. The duration and rate of progression of physical illness determine the psychological reaction of the older person; for example, a rapidly progressing illness (malignancy) frequently leads to more affective disturbance (anxiety, depression) than does a slowly developing condition. Also, the nature of the illness, e.g., whether its effects are visible (skin disease) or invisible (bowel dysfunction), may render the older person more or less disturbed, particularly as regards his or her body image.

As far as the "host" is concerned, premorbid personality and constitutional endowment often come into play in determining the extent of psychological disturbance seen after illness. Congenital defects and/or acquired disabilities such as mental retardation can certainly serve to put the older person at an even greater disadvantage in coping with the limitations of traumatic illness than otherwise would be the case. Similarly, the type of personality involved may determine whether the older person welcomes the disability which may accompany illness or whether the patient is threatened by the effects of the illness on their person and range of functions. Verwoerdt suggests that passive-dependent persons, guilt-ridden or machochistic individuals, and immature or inadequate persons will most often welcome the limitations imposed on them by physical illness because it provides them with opportunities for gratifying chronic unmet dependency needs, satisfies their need for punishment, and helps them to avoid adult responsibilities. Schizoid persons and schizophrenic individuals may also welcome the "sick role" imposed on them by illness in that it provides "a secure bridge leading toward another human being" in the first instance and serves as the "last anchorage point in reality" in the second case (Verwoerdt, 1976, p. 101-102). On the other hand, acting-out personalities, obsessive-compulsive or perfectionistic individuals, and narcissistic persons evidence anxiety, depression and regressive behavior as a result of physical illness. Such individuals because of the limitations of illness can no longer use actions to ward off intrapsychic conflict and tension, can no longer meet the high standards they set for themselves, and they view illness as an attack on their physical attractiveness, strength, and sexual vigor. In all three cases, illness represents a loss (either real or symbolic) and results in feelings of helplessness, hopelessness, despair, and lowered self-esteem.

Situational factors such as the physician-patient relationship

and whether illness requires hospitalization or not are important
in determining the extent of somatopsychic reactions. A poor physician-patient relationship will no doubt exacerbate psychological
problems resulting from illness whereas a good, stable relationship may provide the elderly patient with an element of social support and also encouragement for efforts toward successful coping
and adaptation. Similarly, placing the older person in a hospital
may increase the magnitude of psychologic disturbance (anxiety,
depression) observed secondary to illness, especially if the older
person interprets hospitalization as a confirmation of his or her
worst fears about the severity of their illness, the degree of
pain and suffering they are likely to experience, and the issue of
impending death.

Somatopsychic illness may be more characteristic of the aged
than psychosomatic illness. While the evidence for the latter
among elderly persons is sparse and to some extent contradictory,
it is clear that the risk for physical illness increases with age
and that "Because general homeostatic capacity, recuperative ability, resilience, and energy gradually decline with advancing years,
stresses in late life may have a greater impact than those occurring earlier" (Verwoerdt, 1976, p. 98).

Definition of Related Concepts: Hysteria, Hypochondriasis, and Somatic Delusions

Psychosomatic and somatopsychic disorders seen in the elderly
are often confused with disorders which reflect a somatic expression of psychiatric illness, namely hysteria, hypochondriasis, and
somatic delusions. The latter disorders, despite the fact that
they involve somatic symptoms, are primarily regarded as mental
problems, and thus may be subject to different treatment strategies
than might be employed with psychosomatic illnesses.

Butler and Lewis (1973) suggest the following as signs of
psychosomatic disorders, as differentiated from conversion hysteria
and hypochondriasis: (1) involvement of the autonomic nervous
system, rather than the voluntary musculature and sensory percep-
tive systems; (2) failure of the disorder to reduce anxiety, as
contrasted with a conversion reaction, in which the phenomenon of
la belle indifference occurs; (3) evidence of a physiological
rather than symbolic origin of symptoms; and (4) a definite somatic
threat resulting from structural change.

Conversion hysteria (see earlier case history), on the other
hand, is indicated when a psychological conflict is expressed in
terms of physical symptomatology. The hysterical symptoms usually
involve components of the sensory (hysterical blindness) or motor
(hysterical paralysis) systems, with no evidence of actual physi-

cal impairment. Emotional disturbance is generally absent, i.e., no anxiety or depression despite the loss of function(s), and the somatic symptoms frequently do not correspond with objective, anatomic facts.

Hypochondriasis, according to Busse and Pfeiffer (1969), can be differentiated by the following three mechanisms: (1) a withdrawal of psychic interest from persons and objects in the external environment and a refocusing of this interest on one's self, one's body, and its functioning; (2) a shift in anxiety from specific psychological conflicts to a less threatening concern with bodily disease in general; and (3) the use of physical symptoms as a means of self-punishment and atonement for intolerable feelings (e.g., anger) toward a person close to the individual. Hypochondriasis, unlike hysterical disorders and psychosomatic illness, often involves only excessive somatic complaints and not actual physical symptomatology and impairment. However, it may also occur secondary to chronic physical illness, for example, where an older person with degenerative disc disease eventually complains of aches and pains in several areas of the body including the lower back, legs, shoulders, headaches, and so forth. Busse and Pfeiffer (1969) note that there is a higher prevalence of this disorder in older females, as compared to older males; while Sternbach (1974) suggests that, at least for chronic pain syndromes, older males who manifest hypochrondriacal symptoms do less well in psychological treatment.

Somatic delusions involve an unrealistic belief on the older person's part that he or she is experiencing physical symptoms, which they in point of fact are not. Verwoerdt (1976) briefly describes an example of this relatively rare condition in an 82-year-old woman diagnosed as a paranoid schizophrenic who believed she was 2 1/2 months pregnant. He also points out that this condition is more often associated with psychotic depression, organic brain syndromes, dementia, and schizophrenia than it is with manic and paranoid disorders.

Issues In Assessment

A number of issues present themselves in regard to the assessment of psychosomatic disorders and related conditions in older persons. To begin with, older persons with such disorders will generally not seek out the services of a mental health professional directly. Rather, most elderly clients manifesting some type of mental disorder or somatic symptoms related to psychological factors will first present themselves to a medical specialist, e.g., a general practitioner or internist. The clinical psychologist must, thus, depend on the referring physician in most instances to make appropriate referrals and also to communicate to the older

person that psychological services are essential to a proper diagnostic understanding of their physical problems and the subsequent
treatment thereof. If the physician fails to do this, the psychologist may frequently be faced from the outset with a patient who is
confused about the psychologist's role in the diagnostic process,
who is hostile (as a defense against anxiety and uncertainty) towards the psychologist and who is uncooperative with the various
psychodiagnostic techniques which may be employed. Gentry (1970)
has previously emphasized the necessity of early rapport building
with older clients referred for psychological evaluation in order
to counteract their usual initial reaction of cautiousness and
distrust towards the diagnostician.

A second issue in assessment has to do with the general difficulty of administering standardized objective and projective tests
to older persons in an effort to understand the nature of their
psychosomatic disorders. Gentry (1970) and Bernal, Brannon, Belar,
Lavigne, and Cameron (1977) have both stressed the problem of
"response inhibition," whereby the older patient is motivated more
by a fear of failure than a need to achieve on the psychometric
tests, and where older persons are more affected by fatigue, short
attention span, hearing difficulties, and performance anxiety in
responding to such tests than are younger clients. It is clear
that certain changes in administrative procedures may be required
when testing older patients, e.g., repeating instructions, allowing for frequent rest periods, scheduling the shorter and simpler
tests earlier in the diagnostic process with the more lengthy and
complex tests coming later, and conducting the tests over multiple
sessions. If the diagnostician does not fully appreciate the
physical limitations of the older client, especially those experiencing somatic symptoms, and adjust his relationship with the client so as to facilitate maximum performance on the psychological
tests and openness in the interview situation, much valuable information will be lost and a clear appreciation of the psychosomatic or somatopsychic process will be impossible to achieve.

A third issue has to do with the fact that one may see concurrent organic, physical illness and psychophysiologic/psycho-
somatic illness in the elderly client at the same time. Often it
may be quite difficult to differentiate what is somatic symptomatology secondary to a psychological conflict or stress and what
is simply physical illness resulting from old age. This would
certainly be true for disorders such as bladder and bowel dysfunction and essential hypertension.

Finally, special attention should be given in the assessment
process to those factors which are known to be associated with
psychosomatic disorders in older persons. That is, the diagnostician should focus on issues of loss (loss of significant persons,
job status, material possessions), the process of grief including

anticipatory grief, feelings of despair and depression, the possibility of anniversary reactions, a sense of helplessness or hope-
lessness--all of which characterize the older person who is experiencing psychosomatic disorders. Similarly, premorbid personality
structure, body image, situational factors such as the physician-
patient relationship and response to hospitalization, and the nature
of the illness in question, i.e., rate of progression, severity, and
visible as opposed to invisible symptoms, should be considered when
the diagnostic issue relates to somatopsychic reactions.

REFERENCES

Angle, H. V., Ellinwood, E. H., Hay, W. M., Johnsen, T., and Hay,
    L. R. Computer-aided interviewing in comprehensive behavioral
    assessment. Behavior Therapy, 1977, 8, 747-754.

Bernal, G. A. A., Brannon, L. J., Belar, C., Lavigne, J., and
    Cameron, R. Psychodiagnostics of the elderly. In W. D. Gen-
    try (Ed.), Geropsychology: A model of training and clinical
    service. Cambridge, Mass.: Ballinger, 1977.

Busse, E. W., and Pfeiffer, E. Functional psychiatric disorders in
    old age. In E. W. Busse and E. Pfeiffer (Eds.), Behavior and
    adaptation in late life. Boston: Little, Brown, 1969.

Butler, R. N., and Lewis, M. I. Aging and mental health. St.
    Louis: Mosby, 1973.

Dovenmuehle, R. H., Busse, E. W., and Newman, G. Physical problems
    of older people. In E. Palmore (Ed.), Normal aging. Durham:
    Duke University Press, 1970.

Engelsmann, F., Murphy, H. B. M., Prince, R., Leduc, M., and Demers,
    H. Variations in responses to a symptom check-list by age,
    sex, income, residence and ethnicity. Social Psychiatry,
    1972, 7, 150-156.

Gentry, W. D. The role of the psychologist. In Guidelines for an
    information and counseling service for older persons. Durham:
    Duke University Center for the Study of Aging and Human Devel-
    opment, 1970.

Gentry, W. D. (Ed.). Geropsychology: A model of training and clin-
    ical service. Cambridge, Mass.: Ballinger, 1977.

Lindemann, E. Symptomatology and management of acute grief. In
    A. Monat and R. S. Lazarus (Eds.), Stress and coping: An
    anthology. New York: Columbia University Press, 1977.

Nowlin, J. B., and Busse, E. W. Psychosomatic problems in the
    older person. In E. D. Witkower and H. Warnes (Eds.),
    Psychosomatic medicine. New York: Harper and Row, 1977.

Sternbach, R. A. Pain patients: Traits and treatments. New
    York: Academic Press, 1974.

Verwoerdt, A. Clinical geropsychiatry. Baltimore: Williams and
    Wilkins, 1976.

Weiner, H. Psychobiology and human disease. New York: Elsevier,
    1977.

Whanger, A. D., and Busse, E. W. Geriatrics. In B. B. Wolman
    (Ed.), The therapist's handbook: Treatment methods of mental
    disorders. New York: Van Nostrand Reinhold, 1976.

Section 3
Therapy with the Aged

INTRODUCTION TO SECTION 3:   THERAPY WITH THE AGED

Martha Storandt

Washington University

St. Louis, Missouri

Research on psychological therapies as they are appropriate to older adults is in its infancy. Therapists from various disciplines have largely ignored the needs of this special population. Hence, new or modified treatments have been sparse and investigations of either the effectiveness of specific treatments or the many factors which may influence the appropriateness of their application to older adults are generally lacking. Also, those disciplines which have been faced with the responsibility of caring for the mentally impaired aged in medical and institutional settings are not the disciplines which, by and large, think in terms of control groups, pre- and post-treatment comparisons, or the sophisticated statistical techniques necessary to examine the multidimensionality of the treatment of mental illness in any age group. Other than a few pioneers, psychologists, and especially clinical psychologists, with their specialized training in research methodology and its application to the mental health field are only now beginning to involve themselves in the clinical psychology of aging.

In the preparation of this book the editors were faced with a surplus of need for information and a dearth of information to be found, especially with respect to therapeutic intervention techniques. Only two general classes of treatment techniques were considered sufficiently advanced to warrant separate chapters--- physical treatments primarily involving chemotherapy and the application of behavioristic principles to the treatment of mental disorders in older adults. These two topics are discussed in Chapters 9 and 10, respectively. Other "therapies" which have been used with, or suggested for use with, older adults are included in a chapter entitled "Other Approaches." This pot pourri covers such diverse topics as milieu therapy, reality orientation,

and hyperbaric oxygenation. Not all the possible procedures which
can, or should, be used in the treatment of the mental health problems of older adults are included in that chapter. For example,
day care may be most helpful in some cases. However, such facilities may be equally appropriate to those suffering from mental and
from physical problems. Thus, day care is not the province of
clinical psychology alone and generally is not seen as a "therapy"
but as a resource.

Another area which has been excluded in this section is that
which deals with the application of the principles of community
psychology to older adults. This topic is of such great importance
that it deserves a separate chapter. However, little research
currently exists with respect to this topic; thus, such a review
must wait upon the development of this subspecialty.

Section III begins with a thought-provoking examination of
personality theories as they relate to the treatment of older
adults and a description of the helping relationship as it specifically applies to the therapist with an aged client. Personality
theory as it relates to therapy with older adults has been sorely
neglected in the past, more so than any other topic covered in
this section.

The purpose of this section, and of the book as a whole, is
to provide a concise but thorough review of the literature to
date which relates to the clinical psychology of aging. This section is not necessarily a "how to do it" handbook for the therapist who wishes to serve older clients. Instead it should serve
as a launching platform for students and professionals alike---all
those committed to building and developing a science of therapies
which can be applied to help those older adults who are in psychological pain.

PERSONALITY THEORY, THERAPEUTIC APPROACHES, AND THE ELDERLY CLIENT

Robert Kastenbaum

Cushing Hospital
  and
University of Massachusetts--Boston

How does one person go about helping another? This elementary question may be worth a moment's reflection before we turn to the specific topic of psychotherapy with the elderly. Turning to the problems and potentialities of the elderly is, in fact, a tropism that has attracted relatively few personality theorists and psychotherapists. This means that the newcomer to the clinical psychology of later life cannot count on the guidance available in other areas of therapeutic endeavor. If a person is to work within a coherent and viable perspective, then, this is something that must be constructed by individual effort rather than simply picked up at the check-in point. And so we begin with a brief consideration of the helping process in general and some of the dynamics that have been impeding its development with the elderly client.

HELPING ANOTHER PERSON

The Impulsive and the Systematic

Concentrate, if you will, on the differential between offering and withholding help to another person: why do we sometimes do one, and sometimes the other? By far, most of the help people receive from each other in this world is predicated upon direct human bonds. We recognize that somebody is in difficulty. The recognition quickly recruits the desire or sense of obligation to help. We step forward either with specific help or confidence in our ability to say or do something useful. In this simple paradigm, the flow from recognition to helpful action may be so smooth and immediate that the intervening processes require little reflec-

199

tion. We offer our help in a manner that has the character of an
impulsive, spontaneous action. It is through such "impulsive"
actions, for example, that lost children are reunited with their
parents in a crowded place, or the motorist with a disabled car is
rescued. Psychotherapy? No. Helping behavior? Yes.

However, this sequence is far from automatic. We may not
recognize the other person's distress. Having recognized it, we
may have competing motives that interfere with the inclination to
help (e.g., too busy pursuing our own purposes, fear that we may
be exploited or threatened if we become "involved"). Even with
both the recognition and the inclination, we may not be confident
in our ability to be helpful. ("This person needs a mechanic or
a tow truck; I really don't know a thing about cars.")

Impulsive help probably occurs most frequently and successfully
within a social context in which the general human bonds are strong
and direct. We know each other well enough to recognize signs of
distress; the sense of mutual obligation and basic trust is well
established. Furthermore, there is either little need or expectation of highly technical assistance.

It is no secret that impulsive-empathic help-giving cannot be
entirely relied upon today. There has always been a supplementary
tradition in which care is offered through a more specialized route
(e.g., the shaman, the apothecary). But in recent years there
has been a decided shift toward the technical and institutionalized
mode of help. We turn to physicians, lawyers, mechanics, and the
legions of service-providers in other areas for many of our own
needs; it is not surprising that we expect other people to seek
professional assistance rather than rely upon our layman's ignor-
ance in fields other than our own. In fact, a professional has
sometimes been defined as a person who interprets another person's
crisis as just another routine exercise of his own craft.

The person who offers professional or systematic help does
not invariably respond to need. Examples of non-response are
abundant. The help-giver may see himself as in the wrong specialty
for the situation (or, to put it the other way around, may see the
potential client as having the wrong problem). There is the question of payment. There are the questions of social value and
prestige even if the potential client has the "right" problem.
But when the professional or designated specialist does step forward to provide help, there is the expectation that this service
will derive from a firm knowledge base and technique. The tele-
vision repair person, for example, will not only be knowledgeable
in the general workings but will also have a technical reference
library that enables him to move swiftly and surely to curing
whatever ails your particular set. The physician who scratches
his head and says, "Well, I never saw anything like this before!"

has our attention, but not necessarily our applause. We expect
specialized help to flow smoothly from established knowledge with
a clear sense of goal and technique.

Under ideal circumstances, help for elderly people (whether
of psychotherapeutic or other type) would be readily available
from both the impulsive-empathic and the professional-systematic
resources in society. Let us now remind ourselves just how un-
ideal the circumstances really are.

The Helpless Helper

This time, let us begin with the professional help-giver. We
concentrate upon the potential psychotherapist, although parallels
can be found with other types of service providers.

Certain facts speak for themselves. Journals and books in
the clinical psychology area have given very little attention to
the situation of the older client. Lawton (1970) examined nine
relevant journals over a three year span; of their almost 2000
articles, only 41 had gerontology content (three of the periodi-
cals had no such articles at all). None of the 32 books had adequate coverage, with gerontologic material integrated into the
appropriate content areas. Most of the books read by clinical
psychologists either had no references or one or two brief allu-
sions to the elderly. Reversing perspectives, Lawton was able to
find a few good examples of clinical psychology approaches presented in the field of gerontology, but the general picture was
similar to clinical psychology's neglect of the elderly. He was
forced to conclude that "the mutual interpenetration of gerontological and clinical psychological literature, and therefore,
teaching, approaches the zero level" (Lawton, 1970, p. 154). He
saw the underlying problem as a slightly altered "scientific
version" of our culture's general "ageism" (Butler, 1969). But
whatever the causes of this neglect might be, it is obvious from
Lawton's survey that there has been little in the professional
materials presently available to encourage and guide the clinical
psychologist to concern him/herself with the elderly, or the
gerontologist to focus on clinical issues. The person who happens
to find himself in a potential help-giving situation, then, might
well experience a sense of self-doubt, pessimism, or helplessness.

When the estimated need for psychiatric-type services is
placed alongside the actual amount of service provided for the
elderly, the discrepancy is marked. Kahn (1975), for example,
points out that our nation's mental health system actually serves
a smaller proportion of older persons today than in previous decades. "Rather than helping older persons, the great mental health
revolution has only led to their dropping out of the psychiatric

system" (Kahn, 1975, p. 25). He reports a 45% reduction in the number of patient care episodes for the aged from 1966 to 1971 at psychiatric units in general hospitals. In 1971 the aged in community mental health centers represented only 7% of inpatient service, and only 3% of the outpatient cases.

Both of these latter figures, of course, are well below the proportional representation of older people in the population and, by most estimates, elderly adults tend to have more, rather than fewer, incidents of psychological problems that could benefit by professional care. Kahn is also among a number of observers who express concern over a tendency for an increasing attitude of custodialism in the care of the aged. Butler's (1975) recent overview also makes it difficult to assume that significant progress has been made in providing clinical services to the elderly. He interprets the available data as showing that "therapeutic nihilism" toward the aged is reflected in the excessive incidence of psychopathology in later life.

Intensive studies in specific communities emphasize both the need for psychologically helping services among the elderly and the failure of the mental health system to reach those in need. The rate of what Abrahams and Patterson (1977) term "mental morbidity" in their study of a New England community (17%) is consistent with what at least two other research groups have found elsewhere in the U.S.A. (Berg, Browning, Hill and Wenkert, 1970; Lowenthal, Berkman and associates, 1967). Very few of the distressed and vulnerable elderly were actually receiving any help from mental health specialists.

Psychological help and psychotherapy are not synonymous, of course. How many elderly people are appropriate candidates for psychotherapy in general or for any specific kind of psychotherapy is a question that appears impossible to answer at present. There has been so little basic contact between the vulnerable elderly and the mental health professions that the type of careful assessment from which treatment of choice might be determined has seldom been done. Furthermore, recommending an elderly person for psychotherapy all too often might be an idle exercise, considering the limited availability of psychotherapists willing and able to provide such services.

In attempting to put the "reluctant therapist" (Kastenbaum, 1964) on the couch some years ago, it was easy to recognize the lack of literature and clinical training that Lawton's study subsequently confirmed. But this was just the beginning. The typical psychotherapist appears to represent faithfully the basic discriminatory attitudes rife in our society (e.g., Hollingshead and Redlich, 1958). The low status of the aged in the U.S.A. reduces their attractiveness in the eyes of the therapist. Associ-

ation with a "low-status" patient (whether for age, socioeconomic
or other reasons) lowers the clinician's self-esteem and his
standing with colleagues. One psychiatrist told me bluntly, "If
word gets around town that I see old patients, they'll think I
can't get any other kind." Some psychotherapists also feel "con-
taminated" by the elderly because of the belief that supportive
techniques are the only ones called for--and the concurrent belief
that support is a low-level, unsophisticated procedure not worthy
of gifted therapists. (Both of these assumptions are open to serious question, but this does not subtract from their influence.)

Another set of attitudes centers around the clinician's own
satisfactions and anxieties. There frequently is an expectation
of being "stuck" with a depressed, anguished person who will bur-
den the therapist unbearably. Although therapists expect to be
exposed to the sorrows of their clients, here the insulating distance and the pain-pleasure balance appear to be distinctly unfavorable. Working with a severely psychotic person, for example, can
make great demands. But the therapist has little fear of becoming
psychotic himself; it is possible to maintain a helpful therapeutic distance. This psychic maneuver is more difficult when the
client is aged; professionals in our society appear to be as distressed about the prospect of growing old as anybody else (e.g.,
the graduate students of gerontology studied by Kastenbaum, Derbin,
Sabatini and Artt, 1972). Not only is there a generalized expectation that interaction with the elderly client will result in much
personal distress and little, if any, gratification for the therapist, but there is the special association between the aged and
death to arouse the clinician's own anxieties (Feifel, 1959). The
prospect of a prolonged, intimate encounter with a low-status person for whom one has little clinical or theoretical background to
go by, and whose dysphoria threatens unremitting distress for the
would-be helper while reminding him constantly that he, too, will
grow old and face death--all of this is apt to strike the psychologist or psychiatrist as a less than therapeutic personal experience.

The most critical consideration has yet to be mentioned.
Standing at least knee-deep in the biases of our culture, the
therapist looks at the old person and shakes his head like any
businessman asked to make a bad investment. "Time is money,"
after all. Why invest time and effort in a person with a limited
life expectancy? It is just good sense to help children and young
adults. One's efforts will be repaid through their many years of
better functioning. But the old person will not be around very
long whatever one does. This manifestation of the hedonic calculus
appears to many in our society as self-evident truth. Even the
psychotherapist who is able to transcend the emotional risks expected in intimate encounters with the aged may draw back simply
because he considers his services too important to be wasted on
people with limited futurity.

It is not really necessary now to detail all the cultural dynamics associated with the paucity of help made available to our
elderly; these have been well enough represented in the person of
psychotherapists. Mention might be made, however, of the increasing pressure on the young and middle-aged adult population in this
nation to provide for both the dependent child and the dependent
old person (Shanas and Hauser, 1974). Psychotherapy remains a
relatively high priced and exotic service. It may appear very low
on a total list of priorities for an economy that has not developed adequate support for such obvious necessities as nutrition
and medical care for the elderly. Furthermore, the old suffer
from the general shift from a face-to-face to an institutionalized
type of help-giving. As a group, old people also seem to seek
psychotherapy less often than young adults, for reasons that have
not been determined.

The increased segregation of the elderly from the rest of the
population also must be acknowledged as a contributing factor. It
is not simply that there are more "special places" for old people
to live (and a general increase in the age-segregation of neigh-
borhoods), but the perception that old people are "different" from
the rest of us and should more or less go their own way. This
complex of fact and attitude interferes with the impulsive-spontaneous form of helping behavior. We are not likely to recognize
problems in people who are thought to be much different from ourselves; even less are we likely to help the person we neither see
nor hear.

The potential helping person, in other words, often has
maneuvered himself into helplessness. Low daily interaction with
the elderly reduces the opportunity to recognize and respond to
their problems in a natural, spontaneous way. At the same time,
professionalism in psychotherapy as well as many other service
fields has developed little to guide and inspire systematic helping. This contributes to a self-perpetuating situation: without
successful psychotherapeutic experiences there is little to challenge the nihilism noted by Butler and others. It is not surprising to find that custodialism has become an increasingly common
"solution." Routinized and physicalistic "management" is more
popular than an approach such as psychotherapy that requires human-
to-human contact on an intimate basis. Medication is a frequent
and hasty answer to any distress of the elderly, whether essentially physical, emotional or social in origin. This, of course,
has increased the number of iatrogenic disorders among the elderly
as well as conveyed our intention to keep them at a distance during the helping process. Even a book with such a promising title
as Genesis and Treatment of Psychologic Disorders in the Elderly
(Gershon and Riskin, 1975) proves to be exclusively concerned with
physical interventions. One is reminded all too often of Watson's
observations which reveal an inverse relationship between the

professional echelon of the care-giver and the actual amount of
direct contact with elderly patients (Watson, 1970).

It is hard to escape the general conclusion, then, that the
distressed and vulnerable elderly person cannot count upon a positive response from either the impulsive-spontaneous or professional-
systematic human resources of society. At this point in the discussion it is tempting to preach and advocate. Perhaps a more
useful exercise would be to examine the possibility of successful
psychotherapy with the elderly. Granted that the intellectual and
social climate has not favored much activity in this area—this
does not necessarily mean that psychotherapy would prove to be an
important and effective procedure with the elderly, nor does it
necessarily comprise a mandate for developing all the paraphrenalia
of a specialized clinical field. We have a few questions to ask.

WHAT IS SO SPECIAL ABOUT PSYCHOTHERAPY WITH THE ELDERLY?

The voice of parsimony deserves an immediate hearing. Do we
need to give systematic attention to the elderly client as such?
Age is only one way to classify populations. Gerontologists and
advocates for the elderly often protest that age is relied upon
too heavily by our society in a wide variety of situations. We
might, then, be contributing inadvertently to overemphasis on
chronological age by cultivating this approach to psychotherapy.

There are a number of alternatives. Women predominate among
older Americans, their proportion relative to men increasing the
further along the life-span we move. Furthermore, widowhood is a
common situation. Personality and therapeutic approaches specific
to women or to widows might have more to offer than a broad visu-
alization of "the elderly" per se. Illness and impairment often
accompany the later years of life. Perhaps psychotherapeutic
approaches that are attuned to the person with physical handicaps
and impairments or with so-called "psychosomatic" illness would
be more useful than those that emphasize age as such. Bereavement
and other significant losses are experienced by many elders (in
addition to the specific bereavement of spouse). Again, attention
to a relatively specific circumstance or source of stress might be
more helpful than conjuring with chronological age.

A few more alternatives can be quickly noted. Mandatory retirement continually creates a new crop of the unemployed and
nearly unemployable. The dynamics of forced separation from a
life-long occupation can afflict a person at any point during the
adult years, even if more prevalent in the 60s and 70s. Perhaps
attention to the loss of occupation and the threat of rolelessness
is more important to the theory and practice of psychotherapy than
chronological age. Many elderly are clearly people in the midst

of significant life transitions. This offers still another alternative approach that focuses on people with similar problems and
experiences rather than age as such. The elderly person who is
contending with profound transitional phenomena in his life may be
more similar to younger adults who are also in transition than to
age-peers whose lives are holding steady. Again, we might concentrate upon a predominant distress syndrome rather than age per se.
Many of the elderly men and women who come to the attention of
mental health professionals appear to be depressed--yet depression
is by no means limited to the aged. Theory and therapy centering
around the treatment of depression might therefore be more appropriate than a psychotherapy aimed toward the elderly in general.

        The parsimonious view implied here would have the effect of
discouraging an all-out effort to cultivate a special field of
geriatric psychotherapy. We would examine instead the specific
characteristics of the individual (e.g., an ailing, recently
widowed woman who has sunken into a depressive state), and proceed
on the basis of our best knowledge concerning care of the ill,
the bereaved, and the depressed, as well as whatever we understand
about distinctive dynamics, problems, and strengths of women.

        This approach would also undercut the impression that little
is known about psychotherapy with the elderly. This would be
seen as a misperception based upon overemphasis on age per se, and
underemphasis on therapeutic approaches that are more specific to
the nature of the problem or of the individual. It might also
protect us from formulating theory and techniques that have more
to do with one particular generation of people who have grown old
at one particular time in our social history than with problems
that are intrinsic to aging and the aged. Just as generals have
sometimes trained their armies to fight the last war over again,
so clinicians have sometimes oriented their apprentices to identify
and treat conditions that are in the process of disappearing along
with the particular generation of people who displayed them. Overall, locking ourselves into an approach that establishes geriatric
psychotherapy as a highly specialized field might only perpetuate
age-segregation in theory and knowledge as well as social practice.

            WHAT DO EXISTING PERSONALITY THEORIES HAVE TO OFFER?

Modest Expectations

        The personality theories most familiar to clinical psycholo-
gists do have some implications for psychotherapy with the elderly.
Our expectations, however, should be tempered by the realization
that later life has been only an after thought in the creation of
these conceptual frameworks. What theory of personality ever

started with the intention of explaining aging and old age? What
theory has drawn richly upon the experiences of elderly people in
its initial formulations? What theory even makes old age seem
interesting? The typical theory of personality is fascinated with
early development. Momentum is maintained until midlife at the
most, and then there is progressively less and less to say about
what becomes of us and why. This situation might have been otherwise had Freud been Chinese, or started life as the remarkable old
man he eventually became. As it stands, we are well stocked with
theories that parallel our culture's dominant values. If everybody vanished off the earth a decade or two after the peak parental period of life, this would have only limited impact on most
formulations of personality.

Furthermore, the relationship between theory and therapeutic
procedure is tenuous in general. Professionals such as clinical
psychologists and psychiatrists prefer to believe that (a) we
understand personality, (b) we can help people through the procedure known as psychotherapy, and (c) our theories and our therapies have something to do with each other. But it has proven
very difficult over the years to establish empirically either the
"truth" of our theories or the "success" of our therapies, if we
demand rigorous and systematic evidence on these points. And even
when we have persuaded ourselves that we do understand and are
being helpful, it is still another matter to determine how much of
our actual therapeutic behavior is based firmly upon our theoretical positions. It is possible that the actions of the experienced
therapist depend more upon the pragmatic lessons he or she has
learned than upon abstract theory. While this possibility has
been addressed occasionally in research into the psychotherapeutic
process, a definitive conclusion seems to elude us. "I am reinforcing assertive behavior," one therapist tells himself; "I am
utilizing the transference dynamics," is another therapist's explanation for what might be, in this situation, rather similar
actions. Both may actually be making use of their overall life
experiences and clinical trial-error history to come up with an
approach that seems appropriate to the immediate situation; the
linkage to theory may be only a courtesy reflex.

Similarly the "same" theory may generate either conflicting
approaches to the clinical problem, or no clear approaches at all.
A cognitive theory of personality such as Kelly's (1955) may
embody many admirable structural characteristics. On its own
terms, the theory is laid out with precision and clarity. Nevertheless, whether or not the psychology of personal constructs is
useful for a particular psychotherapeutic challenge may depend
entirely upon the therapist's knack for drawing appropriate inferences. The fact that this approach does not yet seem to have found
application in the clinical psychology of later life is understandable. Kelly's is but one of many theories that have to be "worked

at" before yielding clues for specific clinical actions. In any event, the existence of a theory that has some distinctive strengths does not by itself engender therapeutic models. Even within the more familiar domain of psychoanalytically-oriented theories, clinicians may take many different messages into their therapy sessions after reviewing the same theoretical formulation. We have to keep working at the possible relationship between theory and clinical practice, whether in geriatrics or any place else. Some of us prefer to keep theory and therapy in separate compartments. We like to "travel light," not burdening our therapeutic flexibility with theoretical formulations that do not seem to be quite the right size or shape for the situations that confront us. In this sense, we do not expect much from theory, perhaps just the vague assurance that there are some general principles out there when needed and the residual faith that we are scholars and professionals, not just technicians with a behavioral tool-kit.

There is still another reason for modest expectations, apart from the general neglect of elderly people in personality theory and the undependable link between theory and practice. With few exceptions, existing personality theories share a negative if not nihilistic view of old age. Put it this way: it is difficult to find a major theoretician in psychology or psychiatry who speaks convincingly about the value of being old. Yes, one can discover an oasis here and there, e.g., those few words of Erikson's (1968) that life-span developmentalists crowd around for comfort and, more substantially, several themes cultivated by Jung throughout his writings. But for the most part the theorists display a negative model of old age. Both by what they do and what they don't say about personality and aging, the caution flag is up: proceed at your own risk if you want to wax enthusiastic about therapeutic interventions with the elderly.

Much as we might deplore this attitude so pervasive in the theoretical foundations of our field, the caution might be well taken. Knowledge of aging and the aged has improved substantially since most of the existing theories were formulated. And, despite all the negative stereotypes around, and all the reasons why psychosocial interventions with the elderly might fail, there are many success stories to tell. We need not be bound by yesterday's pessimism; nevertheless, we might be careful in setting our goals and expectations for at least the immediate future.

PERSONALITY THEORIES AND THERAPIES

We will now consider several of the major theories and therapies for their possible usefulness in helping the elderly client. The emphasis will be upon positive contributions of these approaches, although some notes of caution will also be sounded. In addition

to consulting the original theoretical sources, the reader will
also find valuable summaries and interpretations in Rechtschaffen's
(1959) pioneering review, Cath's (1972) examination of psychoanalytic work, Weinberg's (1975) general survey, and Knight's (in
press) exploration of psychotherapy and behavior change with the
non-institutionalized aged. The material presented here is limited not only by the need for brevity, but also by my own outlook
and experiences; others might well select different points for
emphasis.

Psychodynamic Theory and Some of Its Therapies

It is somewhat artificial to separate psychodynamic theory
and therapy from other approaches. All of us have been so influenced by psychoanalytic thought that even the "purest" non-dynamic
theory and the most unFreudian therapist bear some mark of
indebtedness. However, even though the psychodynamic approach to
human experience and behavior has become a significant dimension
of the twentieth century milieu, it remains useful to distinguish
core or traditional approaches from alternative conceptions that
try to go their own way.

Set aside for a moment the not inconsiderable differences
among various psychodynamic views. Let us remind ourselves what
they have in common that might be important in working with
troubled old men and women. Psychoanalytic theory insists that
we take the individual's developmental history very seriously.
If we accept this obligation, then we must be good learners and
listeners. We must learn to see this particular old person as a
distinct individual who has, in his own way, traversed a universal journey through psychosexual stages and balanced as well as he
could the competing demands of inner and outer pressures. The
theory itself encourages a careful, rather scholarly process of
learning about this particular individual. Yet an essentially
disinterested or gerontophobic therapist might slam-bang his way
to conclusions and interventions with the conviction that he really
is using psychoanalytic theory. Most likely, he is drawing the
most automatic inferences that pertain to "text book" suppositions
of narcissism, regression, ego decline and whatnot without dili-
gent attention to the particular individual who stands (or lies)
before him. If we took psychoanalytic theory itself seriously,
then it would provide at the least the expectation that human
nature is complex, and deserving of our most sensitive observations. The fact that some people use psychoanalytic propositions
as a way of hastily "managing" the elderly reflects mostly on
therapist, not theory.

Psychodynamic theory has often focused special attention on
both the beginning and the termination of therapy. Specific

thoughts on these topics are scattered throughout many published case histories, and upgathered in numerous theoretical papers and secondary sources. It seems that every potential psychotherapist with the elderly would want to become familiar with both the theory and technique here. We have already seen that frequently there is <u>no</u> beginning to therapy with the elderly. Possibly there would be more beginnings and more successful continuations if the therapist had a firm conception of what he should expect from the client and himself--and was able to share this conception with the client. A person sensitive to the psychodynamics of beginnings is better able to avoid initial contacts that court failure by implicit seduction, threat, or untenable premises. Similarly, the problem of terminating therapy has been widely discussed among psychoanalysts. Familiarity with the theory and practice of saying good-bye is vital for a therapist whose client has a very limited life expectancy. This is not to say that psychodynamic theory has easy answers for coping with the entry and exit points of therapy; rather, this rich body of conceptualization and clinical experience confronts us clearly with the need to <u>think</u> about what we are doing all the way through the process. Whether or not a particular therapist operates within a psychoanalytic framework, he can probably profit from learning what others have learned through decades of experience.

Take, for example, one type of decision a therapist has to make before starting to work with an elderly person. Is this client to be treated essentially as an intact adult beset with difficulties that he himself wants to alleviate? If so, the therapist might invoke the famous "fundamental rule" of psychoanalytic treatment: "He is to tell us not only what he can say intentionally and willingly . . . but everything . . . that his self-observation yields him, everything that comes into his head, even if it is <u>disagreeable</u> for him to say it, even if it seems to him <u>unimportant</u> or actually <u>nonsensical</u>" (Freud, 1964, p. 141). The therapist who views old people in general as adults, and this particular old person as an intact adult, would then be able to initiate treatment much as he would with any other individual.

If the therapist believes, however, that old people typically are "regressed" or "damaged goods," or that this particular individual has organic complications that are inconsistent with the ego functioning required by the fundamental rule, then some variation of the start-up rule is indicated. "I want to help this person keep regressive forces under control; I don't want to encourage undisciplined, primary process thought," the therapist might decide. With this alternative outlook, the therapist might initiate treatment by taking a more active role and "setting limits." Instead of the free-wheeling fundamental rule, then, we might have something that approaches therapist-as-parent/client-as-child relationship.

These two are not the only alternatives, of course. But the point is that the psychodynamically-attuned therapist will have made at least a provisional decision at the outset about the over- all goals and course of the relationship, and the level and style at which it most appropriately can be contacted. Whatever his particular decision might be in a given instance, he will be look- ing for the creation of a <u>therapeutic alliance</u>. This is a more contemporary and flexible descendant of earlier therapeutic modes that relied heavily on the doctor's authority and the patient's docility and suggestibility. It acknowledges that there is more than "transference" in the therapist-client relationship, but also an authentic mutual willingness to work together. Again, whether or not a therapist owes principle allegiance to psychodynamic theory, it is important to establish and cultivate an alliance that will see them both through the difficult process of psycho- therapy.

Psychoanalytic theory also sensitizes us to the complex inter- action between what might be called the person's "developmental position" and his characteristic ways of solving or fending off life's problems, the "defense mechanisms." This can provide use- ful insights to the therapist. Consider, for example, the old man or woman who regularly complains "They're stealing from me!" Now, it is indeed possible that somebody is stealing something from them, and it is also possible that, in an institutional setting, another person is carrying off one of their possessions in honest confusion. But it is also possible that being-stolen-from is an explicit, if symbolicized, expression of how the person interprets his overall life situation. He is feeling depleted. Inner re- sources are draining away. Libido is at low tide; the body is not working as it did. Concerns and satisfactions are ebbing back to earlier and more "primitive" psychosexual stages (e.g., bowel move- ments become more interesting than orgasms). Behaviors that seem to be directed entirely to concrete daily life events may be seen as defensive strategies to protect against the perception of regres- sive changes and ill-defined threats from both inside and outside the ego. The defenses themselves may take on the character of earlier psychosexual stages, e.g., become "anal-retentive," or, more primitively yet, "oral-grasping."

Stay with this example a moment longer, if you will. What use might the therapist make of his conclusion that this elderly person's skirmishes and discomforts are based largely upon a fundamental drift toward "regression"?[1] The most traditional

---

[1]"Regression" has taken on a variety of meanings in psychoanalytic theory, as have other key terms. Here reference is made to a re- turn to a simpler, less mature organization of personality as indexed by the so-called psychosexual stages of development.

option would be to work toward insight and, if the ego of both
patient and therapist will bear it, toward a major "reconstruction"
of the personality. It is more likely, however, that the therapist
will take some other kind of action in the case of an elderly person, especially if illness, senile qualities, and unfavorable environmental circumstances are part of the scene. He might, for example, decide that this person needs "permission" to regress a
little. It is all right to be more dependent now; one does not
have to engage in an all-out fight to maintain the stance of a
healthy young phallic or genital stage adult. This decision might
be transformed not only into a particular kind of therapeutic attitude, but also into some actions in the environment itself to help
the person to place trust in the world. With a somewhat different
emphasis, the therapist might try to help the client develop a
sense of replenishment: there is something coming in, too, some
source of "narcissistic supply" or "emotional fuel." The therapist
could try to provide some of this supply himself, thereby giving
the treatment a relationship therapy direction, or could try to
help the old person maneuver more successfully in the environment
to get a little something back.

Here we see one of the significant changes in emphasis from
early psychoanalytic theory to the kind of therapeutic interventions increasingly favored today and which may be especially relevant to work with the elderly client. In _echt_ Viennese psychoanalytic orthodoxy, the "real world" was of secondary interest.
It was what happened between analyst and analysand, and in the
psyche of the latter, that mattered. Improvement in functioning
in daily life would be a consequence of successful therapy. Today, however, it is far more common to use psychoanalytic theory
as one of the bases for a broader scaled approach to treatment,
rather than limited to the direct therapy session. The psycho-
analytically-trained therapist who has taken a sustained interest
in the well-being of aged people is not bashful about suggesting
practical interventions. The guide to understanding comes from a
theoretical perspective, but the solution may well include social,
environmental, nutritional, medical, and even financial interventions. The fact that there has been a steady growth of an ego
psychology approach within the psychodynamic theories and therapies provides a more resourceful and optimistic basis for work
with the elderly. Some illustrations of psychoanalysts addressing
themselves to the integration of theoretical/therapeutic and pragmatic solutions to the psychologic problems of elderly people can
be found in three books emanating from The Boston Society for
Gerontological Psychiatry, Inc. (Zinberg and Kaufman, 1963; Berezin and Cath, 1965; Levin and Kahana, 1967), as well as on the
pages of their _Journal of Geriatric Psychiatry_. Although the individual contributions vary in quality, they all bear the mark of
concerned therapists attempting to find something of value in their
conceptual tradition while also taking cognizance of the "real

world" in which many troubled elderly people live.

From psychoanalytic theory we can also learn the importance of identifying and then respecting the client's defense mechanisms. The old person shows anger, or withdraws from the therapist, let us say. A "natural" response of the therapist would be to get angry in return, or reduce his own involvement in the relationship. Responses of this type most likely are what other people in the environment have been giving to the same provocation from the client. The so-called therapist, in this instance, would have behaved in a nondistinctive and nonhelping way. Familiarity with psychoanalytic theory and practice would alert the therapist to the processes likely at work. Anger, withdrawal and other "negative" actions would be seen as coping techniques that this person finds it necessary to use. The therapist would attempt to discover what needs these maneuvers were attempting to serve. He would not just become another person returning misery for misery. Just the recognition that certain behaviors may have significant defensive functions would enable the therapist to be helpful. He might be the only individual in the old person's life at this time who remains untrapped by the pattern of deepening interpersonal distress. He appreciates that the behavior is not a simple reflection of "orneriness" or "senility," but rather is an effort to guard against an inner peril (catastrophic anxiety? severe depression?). In this sense, the psychodynamically-attuned therapist is an ally in the old person's struggle for continued integrity of self.

This perception of the old person as an individual whose seemingly maladaptive behaviors represent an attempt to fend off even more disturbing psychic states often will be taken as a caution against attacking the defenses as such. Here is a place where those with a general lack of therapeutic experience or a superficial interpretation of psychoanalytic theory are likely to go wrong. "There's a defense! Attack!" The mature therapist working with elderly people recognizes the dangers in attempting to take away the few remaining coping techniques that have been pressed into service. (Fortunately, many elders are adept in turning aside ill-advised onslaughts on their defenses.)

Verwoerdt (1976) offers good advice in utilizing a psychodynamic approach to elderly clients. He emphasizes that "To remove defenses can be compared with taking away crutches from a paralyzed patient. When a particular defense is removed, another one is likely to take its place" (Verwoerdt, 1976, p. 143). Two basic therapeutic strategies are recommended: (a) discover and attempt to alleviate the source of the anxiety which fuels the maladaptive defenses, and (b) help the client establish another type of defense that will be more successful.

This advice, of course, is applicable to therapeutic efforts

with many other people besides the elderly. It may be of particular significance here, however, because the elderly man or woman
is more likely to be bereft of other resources to support self-
esteem and overall functioning (e.g., sick rather than healthy,
bereaved rather than surrounded by loving intimates, confronting
a threatening future rather than one which promises to improve
with time).

One particular source of distress and its expression in "defense mechanisms" may be especially prevalent with the elderly.
This first came to the attention of my colleagues and myself during clinical and research activities with a hospitalized population. The typical situation involved a person with one or more
physical conditions that were considered chronic and essentially
unamenable to aggressive treatment. Nevertheless, effective and
dedicated action by the allied health staff of a geriatric hospital would alleviate the problem, e.g., the old man could walk
again; the old woman had enough bodily control and stamina to live
a more normal type of life.

Paradoxically, or so it seemed, successful alleviation of a
disabling physical problem would be followed by an increase in
depression and agitation. Sustained therapeutic contact led to
the strong impression that removal of the symptom had also removed
the old person's explanation for what had gone wrong with his life.
It had proven easier to focus all the general life distress
around the symptom than to come to terms with "old age" as such.
In sum, he could not accept himself as an "old" person; this would
have been a damaging onslaught on the identity built up over the
years; but it was acceptable to see one's self as a sick or disabled person. Successful treatment of an obvious physical problem
still left the individual "old." This seemed to lead to what
might be called a "crisis of explanation" (Kastenbaum, 1964)
which, fortunately, could be approached therapeutically with some
success.

Goldfarb (1954, 1955, 1956, 1969) has pioneered an application
of psychodynamic theory to therapy with the impaired elderly that
emphasizes the strengthening and refinement of their defenses. He
encourages the vulnerable and maladapting older adult to draw upon
the personal strength of the therapist and, through this relationship, teaches him more effective strategies for meeting his emotional needs. (While Goldfarb's approach is best understood within
the psychodynamic framework, a person with a behavior modification
approach would find it easy to interpret many of his actions in
terms of differential reinforcement and related concepts.) Goldfarb views the old person's need to satisfy dependency cravings as
one of the central factors. In our society it is socially desirable to present one's self as an autonomous being. Throughout
much of adult life, then, a person may more or less masquerade as

an independent individual, holding his own dependency needs in
check or under cover. These needs emerge more forcefully in old
age, however, and the techniques used to force or lure others to
"take care" often are counterproductive. The person might also be
caught in a situation where vestiges of self-esteem require him
to deny dependency needs at the same time that the needs themselves
have become more urgent. Goldfarb advises that the therapist accept these needs as normal and work toward helping the vulnerable
elder meet them more effectively and without driving other people
away.

Rechtschaffen (1959) was critical of this approach and, indeed, it can produce difficulties of its own. Yet one should
bear in mind that the typical old person'that Goldfarb discusses
is one with multiple psychological and organic problems, the sort
of individual who most therapists would turn away from. The value
of his approach would not necessarily be the same for, let us say,
a mentally alert octogenarian who is facing a situational crisis.
However, the attention to dependency needs and how to meet them
within a pragmatically-oriented psychodynamic approach seems
worth considering in a large variety of therapeutic interactions
with elderly people.

We have not taken much notice here of the appreciable differences in emphases, even in fundamental concepts, among psychoanalytic thinkers. It might be worthwhile to examine the work of all
the seminal psychoanalysts for possible applications to therapy
with the elderly. This cannot be done here, but a few illustrations might be useful.

Rank differs from most of the other early psychoanalytic theorists in his emphasis on the selfhood of the individual. With
Freud, for example, we are encouraged to emphasize the <u>analytic</u>
component of psychoanalysis. Rank found it more appropriate to
focus on the person acting as a total being. The difference between a person who is, in effect, hiding from life and the person
who is moving forward can be understood in terms of his "will-
orientation" (Rank, 1950). Psychotherapy is less concerned with
taking apart and re-doing psychic components than it is with helping the person to orient his total being toward a life-affirming
position. He therefore gives attention to such questions as the
competing and mutually complementary use of love and force, and
does not hesitate to use such expressions as "ethical self determination" in discussing desirable therapeutic outcomes. In general,
the Rankian approach may be more congenial than the orthodox
Freudian for those with a more holistic orientation. The individual tends to be portrayed more as a (potentially) free agent than
as a beseiged go-between for conflicting inner impulses and external demands. This difference in outlook has important implications for the role and actions of the therapist. He criticizes the

traditional analyst who "himself becomes the parental love and
force authority, who pretends to remodel the patient in terms of
the educational morality, and the patient accepts these preten-
sions when he raises the analyst to a god and claims him as love
object." By contrast, Rank speaks of letting the individual
"create for himself his own development and his own freeing which he
does anyway but must deny as an expression of his own will as
long as the other remains a symbol of love or force (authority)
. . . ." (Rank, 1950, pp. 67-68).

This approach may have particular relevance to work with the
elderly, especially for the therapist who is not comfortable with
the idea of fashioning himself into a parent-figure for a person
literally old enough to be his own parent or grandparent. Perhaps
the Rankian approach also makes it somewhat easier to develop respect for the individual, especially for the potential "growing
edge" of his personality even in advanced old age (a topic that
Rank does not seem to have addressed in much detail). Rank also
helps us to see crisis situations as opportunities--opportunities
for further growth in personality for which the psychotherapist
might serve as a kind of midwife. However, the general body of
Rankian literature, theoretical and applied, is relatively small
when compared with the Freudian, and it is difficult in practice
to keep this approach clearly distinct from the more traditional
psychoanalytic view and to reconcile differences. Nevertheless,
it is probable that many theorists and therapists would find
Rank's view of personality a useful one to guide their efforts.

It is to Jung, however, that one turns to discover a major
psychoanalytic thinker who has something compelling to say about
the second half of life. Jung's view of human development extended throughout the life-span. Old age is not something as
simple as regression or a pale reflection of what has gone before.
Instead, Jung sees fundamental processes at work that, when carried out fully, transform and complete the individual. His view
is expressed in such statements as: "Nobody seems to consider
that not being <u>able</u> to grow old is precisely as absurd as not
being able to outgrow child-sized shoes. A still infantile man
of thirty is surely to be deplored, but a youthful septuagenar-
ian--isn't that delightful? And yet <u>both</u> are perverse, lacking
in style, psychological monstrosities" (Jung, 1959, p. 6).

Jungian theory emphasizes differential life pathways for men
and women. Essentially, men begin to allow more of their underlying femininity to surface, while women develop the more mascu-
line aspects. These are normal changes in Jung's view, and represent a larger process through which the aging person lays claim to
those potentials of their personality which were not utilized
earlier in life. What it means to become a self receives much more
attention from Jung than from most other theorists, in or out of

the psychoanalytic domain. In reading Jung one often has the impression of a sort of psychic quest in which the individual is trying to reach down to the deepest roots of biological existence and yet upward to the highest flowering of distinctively human characteristics. There is a concomitant fascination with inner life that is, if anything, even more marked than with the Freudians, although differing in many respects. Those who are made uncomfortable by delvings into "the unconscious" and by the interpretation of symbols are not likely to find Jung's realm a comfortable one. Behavior modifiers have to read Jung in the closet.

Whether or not one agrees with Jung's overall approach to personality, it is a rich and fairly distinctive one. There is the signal advantage that Jung actually does give serious consideration to old age, rather than requiring the gerontologist or geriatric clinician to draw all the implications himself. It is also especially valuable, I think, for sensitizing the therapist to the old person's inner life (and to his own as well). Jung's work is perhaps more difficult to comprehend than most other psychoanalysts. This is partly because of the concepts themselves and the writing style, partly because he did not often pause to summarize his observations in concise form. For this reason, it is useful to make friends with several good formulations of his work by other hands (Jacobi, 1962; Fordham, 1966; Whitmont, 1973). Among Jung's own works, there are several that make useful beginnings (Jung, 1959b, 1963, 1973).

Other Theories and Therapies

The popularity and influence of psychodynamic theories should not lead us to neglect other approaches. Of these, the various concepts and techniques subsumed under the rubric of behavioral approaches are reviewed elsewhere in this book (Chapter 10). We will also bypass group psychotherapy, not in any spirit of disinterest, but because the purposes and methods here depend just as much upon the therapist's theoretical orientation as is the case with individual treatment. Discussion of group therapy with the aged has been offered by several clinician-writers (e.g., Goldfarb, 1971; Saul and Saul, 1974; Ross, 1975).

Gerontology's first psychosocial theory of its very own has not given rise to clear implications for psychotherapy. The authors of disengagement theory (Cumming and Henry, 1961) have tried to stay away from advocating particular programs based upon their work. This has not stopped others, of course, from drawing a variety of implications. These have divided themselves roughly into those who believe the disengagement process should be facilitated because it is said to be "normal," and those who believe

that a person is happier when active and engaged. Research findings have not been kind to the basic structure of disengagement
theory as a general description or semi-explanation of what happens
to the individual with advancing age. Whether it is for this or
other reasons, we have not seen the formulation of psychotherapy
approaches that draw directly from disengagement theory. It is
likely, however, that the issues brought into focus by Cumming and
Henry (e.g., complementary or antagonistic relationship between
the expectations of the aging individual and society, possible differences in the fate of instrumental and socio-emotional styles of
life adaptation) will continue to be of interest to all who work
with older adults in our society.

Let us briefly consider a few approaches that might prove to
have useful applications with the elderly. Cognitive approaches
to personality have led, naturally enough, to cognitive approaches
to psychotherapy. We might take as example the previously cited
work of Kelly (1955). Here is a theory of personality that centers
around the way in which the individual construes his world. At
this general level, it has something important in common with a
number of other approaches, including the client-centered outlook
of Rogers (1959). However, Kelly's approach is decidedly mental-
istic. The basic model is that of the human as a rational, logical being--almost a novelty in this day and age! Each person attempts to control the world around him. This necessarily requires
the ability to predict which, in turn, requires some basic comprehension of what the world is about. Essentially, this theory
depicts Everyman as a scientist (although some fit the role better
than others). We try to predict and control by interpreting our
world in terms of a set of ideas or constructs. These are organized hierarchically for each individual: certain concepts are
very broad and are used in a variety of situations; others are
quite specific. One of the most important characteristics of
these constructs is their bipolarity. In other words, if "being a
big success" is a major construct for a particular individual, it
is a sure thing that "being a terrible failure" is also represented in his construct system even if we seldom see it come to
the surface.

It is challenging to think how such a cognitive approach might
be applied to psychotherapy with the elderly. One of the more obvious uses would be to learn how this particular client interprets
his life experiences. The therapist could then determine what
"wave-length" transmits and receives. In general, this approach
might provide some fresh options in therapeutic situations where
emphasis on emotional and dynamic factors does not seem to be making headway. It is also likely that a therapist who approaches an
elderly client as though a rational, truth-seeking person would be
communicating a distinctive attitude and set of expectations that
could in itself be valuable. On the negative side, perhaps the

most evident problem would be a lack of sensitivity to the old person's feeling tone and the finer emotional nuances of the situation. Cognitive therapy with older people might be an especially useful comparative treatment technique if one were interested in evaluating the merits of various approaches empirically.

The study of "learned helplessness" has yielded an interesting array of experimental findings with animals, and a number of possible parallels to the human condition. The work of Seligman and his colleagues might not be what one usually has in mind when speaking of personality theory, but it could prove relevant to our purposes. In a sense, this is another manifestation of cognitive theory, although it has a surface resemblance to stimulus-response theory. Many of the studies are reviewed in a recent paper by Maier and Seligman (1976), while the broadest presentation can be found in Helplessness (Seligman, 1975). The subtitle of this outstanding book indicates its relevance to helping the elderly person, although Seligman himself has not been identified with the field of gerontology: On Depression, Development and Death.

Basic to a "learned helplessness" situation is the individual (animal or human) being trapped in an environment that either extrudes noxious stimulation or poses an apparently unsolvable problem. The individual cannot see a way of changing or getting out of the situation, of exercising effective influence over it. This leads to a profound syndrome of what might be called unconditional surrender. Even if an opportunity subsequently arises in which the individual could leave the situation or solve the problem, the attitude of helplessness prevails. A dog might fail to walk through the now-open door, for example, and leave behind a cage that has brought him nothing but misery.

Seligman observes that humans caught in highly institutionalized situations over a period of time may behave in similar fashion. In effect, helplessness is no longer simply the reaction to an intractable situation; it has become the individual's primary response to the world in general. Identification of the "learned helplessness" dynamics could lead to prevention, as well as to alleviation through a variety of modalities. Psychotherapy, either in familiar or innovative forms, would be one reasonable approach to un-learning the state of deep helplessness.

It is possible that other terms that are applicable to the situation of some elderly people--powerless, alienated, useless-- represent much the same condition. Specific therapeutic techniques could be adapted from methods used to try to rehabilitate subhuman creatures who have undergone learned helplessness experiences, or we could bring to bear techniques usually associated with other therapeutic approaches. To the extent that old people suffer because they no longer feel there is any point in trying to

help improve their situation, then to that extent the work of
Seligman and his colleagues might be very relevant.

The work of Carl Rogers has already been mentioned in passing.
It should have a number of appealing aspects for potential therapists with the elderly. This approach does not have its roots in
conventional psychopathology; it is an expression of a philosophy
of relationship between people rather than the outcome of a traditional diagnostic-treatment orientation. As a corollary, it does
not urge the therapist to come across as a highly professionalized
or scientized "know it all." Many people seem to have become competent therapists within this new tradition without extensive educational backgrounds, a feature that greatly increases the number
of potential therapists available. This has also provided a
basis for controversy and criticism, it should be noted. In a
recent statement of his approach, Rogers declares that the central
hypothesis of the client-centered approach is that the individual's
growth potential can be released "in a relationship in which one
person (usually thought of as the helping person) is experiencing
and communicating his own realness, his caring, and a deeply sensitive nonjudgmental understanding. This approach is unique in
being oriented to the process of the relationship, rather than to
the symptoms or their cure . . ." (Rogers, 1975, p. 1832).

The "nonjudgmental understanding" is important to underscore.
The therapist or helping person avoids assuming the role of judge
or moral authority. This approach makes it clear that the therapist is utilizing chiefly his own person in any help that is provided. If being a troubled old person has something to do with a
deficit in honest human relationships, then the client-centered
approach has much to recommend itself. Furthermore, it would
seem to support the client without recourse to some of the words
and actions often associated with the concept of "support." The
very fact that the helping person is there, all eyes and ears,
paying full attention to the client, is perhaps more supportive
(and more appropriately so) than any number of doled-out statements of comfort.

Client-centered psychotherapy has entered the consciousness
of our times, although probably not to the extent of psychoanalytic theory. We were not issuing cliches about people being
"real" or "authentic" before Rogers' client-centered approach came
along—and these are not cliche statements in themselves although
repeated use sometimes makes them seem so.

One extra advantage of client-centered psychotherapy is its
self-monitoring tradition. Resembling behavior modification in
this respect if in few others, there is a readiness to open the
therapeutic process to empirical study. Considering how little we
know about the effectiveness of various psychotherapeutic tech-

niques with the elderly, this is all to the good. On the caution-
ary side, it seems to me that there is a particular danger in
the deceptive simplicity of the client-centered approach. It can
be made to look so easy! But, like any other type of psychotherapy, this approach makes considerable intellectual and emotional
demands on the therapist. The person who decides to use a client-
centered approach with the elderly simply because it looks to be
the easiest method would do well to reconsider.

Numerous other theories and therapies might be examined for
their possible application with elderly clients. There is gestalt
therapy, for example, with its emphasis on expressing one's self
and living in the moment--while existential therapy attempts to
help the individual develop a "meaning" in life that goes far beyond the moment and the self. Which of these approaches, if
either, is appropriate for the old man or woman who is beset with
many problems in the immediate situation and at the same time is
trying to come to terms with the total significance of his/her
life? Or are we asking the wrong question? Should we be asking,
instead, what theory and technique does a therapist feel comfortable in bringing to a helping interaction with an elderly person?
It might be fanciful to suppose that a potential therapist would
select an approach entirely on the basis of the client's characteristics. At the opposite extreme would be the attitude: "This
is the kind of therapy I know how to do: so I will do it with
everybody who comes my way."

CONCLUDING NOTE

Ideally, there would be another section at this point: a review of the studies seeking to demonstrate the effectiveness of
psychotherapy with the elderly. But to have a sound basis for
evaluating the success of various types of therapy with various
types of elderly clients one must, at a minimum, have a variety of
clearly reported therapies to review. The need for good research
into the psychotherapeutic process with elderly people is obvi-
ous--but even more evident is the need for more psychotherapy! In
lieu of a substantial body of psychotherapy research with the
elderly, at least it can be said that pessimism and nihilism have
not been supported and that many types of theory and technique
have yet to be applied systematically. Considering the impressive
individual differences among elderly people and the broad range of
psychotherapeutic approaches available, we have hardly begun to
match treatment modality to the needs and strengths of the individual.

It would be a grim commentary on our culture in general, as
well as on the art of psychotherapy in particular, if we proved

unable to help each other through that most elementary bond, the
human relationship. And it would be, if possible, an even more
sorrowful note, if we didn't bother to try.

REFERENCES

Abrahams, R., and Patterson, R. D. Psychological distress among
    the community elderly: Prevalence, characteristics and im-
    plications for service. International Journal of Aging and
    Human Development, 1977, in press.
Berezin, M. A., and Cath, S. H. (Eds.). Geriatric psychiatry:
    Grief, loss, and emotional disorders in the aging process.
    New York: International Universities Press, 1965.
Berg, R. L., Browning, F. E., Hill, J. G., and Wenkert, W. Assess-
    ing the health care needs of the aged. Health Services Re-
    search, 1970, 3, 36-59.
Butler, R. N. Age-ism. Another form of bigotry. Gerontologist,
    1969, 9, 243-246.
Butler, R. N. Psychiatry and the elderly: An overview. Ameri-
    can Journal of Psychiatry, 1975, 132, 893-900.
Cath, S. H. Psychoanalytic viewpoints on aging--an historical
    survey. In D. P. Kent, R. Kastenbaum and S. Sherwood (Eds.),
    Research, action and planning for the elderly. New York:
    Behavioral Publications, 1972.
Cumming, E., and Henry, W. E. Growing old. New York: Basic
    Books, 1961.
Erikson, E. H. Identity: Youth and crisis. New York: W. W.
    Norton & Co., 1968.
Feifel, H. (Ed.). The meaning of death. New York: McGraw-Hill,
    1959.
Fordham, F. An introduction to Jung's psychology. Baltimore, Md.:
    Penguin Books, 1966.
Freud, S. An outline of psychoanalysis. In Standard Edition of
    the Complete Psychological Works of Sigmund Freud, Vol. 23.
    London: Hogart Press, 1964.
Gershon, S., and Raskin, A. (Eds.). Genesis and treatment of
    psychological disorders in the elderly. New York: Raven
    Press, 1975.
Goldfarb, A. I. Psychotherapy of aged persons. IV. One aspect
    of the psychodynamics of the therapeutic situation with aged
    patients. Psychoanalytic Review, 1955, 42, 180-187.
Goldfarb, A. I. The rationale for psychotherapy with older per-
    sons. American Journal of Medical Science, 1956, 232, 181-
    185.
Goldfarb, A. I. The psychodynamics of dependency and the search
    for aid. In R. A. Kalish (Ed.), The dependencies of old
    people. Ann Arbor, Mich.: Institute of Gerontology, 1969.
Goldfarb, A. I. Group therapy with the old and aged. In H. I.
    Kaplan and B. J. Sadock (Eds.), Comprehensive group psycho-

therapy. Baltimore, Md.: Williams & Wilkins, 1971.

Goldfarb, A. I., and Sheps, J. Psychotherapy of the aged. III. Brief therapy of interrelated psychological and somatic disorders. Psychosomatic Medicine, 1954, 16, 209-219.

Hollingshead, A. B., and Redlich, F. C. Social class and mental illness. New York: John Wiley, 1958.

Jacobi, J. The psychology of C. G. Jung. New Haven, Conn.: Yale University Press, 1962.

Jung, C. G. The soul and death. In H. Feifel (Ed.), The meaning of death. New York: McGraw-Hill, 1959a.

Jung, C. G. The basic writings of C. G. Jung. New York: Modern Library, 1959b.

Jung, C. G. Memories, dreams, reflections. New York: Random House, 1963.

Jung, C. G. Psychological reflections: A new anthology of his writings, 1905-1961. Princeton, N.J.: Bollingen, Princeton University Press, 1973.

Kahn, R. L. The mental health system and the future aged. Gerontologist, 1975, 15, 24-31.

Kastenbaum, R. The reluctant therapist. In R. Kastenbaum (Ed.), New thoughts on old age. New York: Springer, 1964a.

Kastenbaum, R. The crisis of explanation. In R. Kastenbaum (Ed.), New thoughts on old age. New York: Springer, 1964b.

Kastenbaum, R., Derbin, V., Sabatini, P., and Artt, S. The ages of me: Toward personal and interpersonal definitions of functional aging. International Journal of Aging and Human Development, 1972, 3, 197-212.

Kelly, G. A. The psychology of personal constructs. New York: W. W. Norton & Co., 1955.

Knight, R. Psychotherapy and behavior change with the non-institutionalized aged. International Journal of Aging and Human Development, in press.

Lawton, M. P. Gerontology in clinical psychology, and vice-versa. International Journal of Aging and Human Development, 1970, 1, 147-160.

Levin, S., and Kahana, R. J. (Eds.). Psychodynamic studies on aging: Creativity, reminiscing and dying. New York: International Univ. Press, 1967.

Lowenthal, M. F., Berkman, P. L., and associates. Aging and mental disorder in San Francisco. San Francisco: Jossey-Bass, 1967.

Maier, S. F., and Seligman, M. E. P. Learned helplessness: Theory and evidence. Journal of Experimental Psychology, 1976, 105, 3-46.

Rank, O. Will therapy and truth and reality. New York: Alfred A. Knopf, 1950.

Rechtschaffen, A. Psychotherapy with geriatric patients: A review of the literature. Journal of Gerontology, 1959, 14, 73-84.

Rogers, C. A theory of therapy, personality, and interpersonal relationships, as developed in the client-centered framework.

In S. Koch (Ed.), Psychology: A study of a science, Vol. 3.
New York: McGraw-Hill, 1959.

Rogers, C. Client-centered psychotherapy. In A. M. Freedman, H.
I. Kaplan and B. J. Sadock (Eds.), Comprehensive textbook of
psychiatry. II. Baltimore: Williams & Wilkins, 1975.

Ross, M. Community geriatric group therapies: A comprehensive
review. In M. Rosenbaum and M. M. Berger (Eds.), Group
psychotherapy and group function. New York: Basic Books,
1975.

Saul, S., and Saul, S. Group paychotherapy in a proprietory nursing home. Gerontologist, 1974, 14, 446-459.

Seligman, M. E. P. Helplessness. On depression, development, and
death. San Francisco: W. H. Freeman, 1975.

Shanas, E., and Hauser, P. M. Zero population growth and the family life of old people. Journal of the Psychological Study
of Social Issues, 1974, 30, 79-92.

Verwoerdt, A. Clinical geropsychiatry. Baltimore: Williams &
Wilkins, 1975.

Watson, W. H. Body image and staff-to-resident deportment in a
home for the aged. International Journal of Aging and Human
Development, 1970, 1, 345-360.

Weinberg, J. Geriatric psychiatry. In A. M. Freedman, H. I.
Kaplan, and B. J. Sadock (Eds.), Comprehensive textbook of
psychiatry. II. Baltimore: Williams & Wilkins, 1975.

Whitmont, E. C. The symbolic quest. Basic concepts of analytic
psychology. New York: Harper & Row, 1973.

Zinberg, N. E., and Kaufman, I. (Eds.). Normal psychology of the
aging process. New York: International Universities Press,
1963.

ORGANIC TREATMENT OF THE ELDERLY[1]

Philip L. Kapnick

St. Louis College of Pharmacy

Although clinical gerontology is a relatively new area of specialization, psychologists who practice in this field are committed to treatment goals which are similar to other areas of clinical psychology. Therapeutic goals, in a very general and ideal sense, are: to minimize the patient's suffering, to improve behavior, to lessen interpersonal friction, and to assist the patient in taking pleasurable interest in his or her environment. To achieve these goals a combination of methods may be necessary. One method which is often used is psychopharmacology, the treatment of the emotionally disturbed with psychotropic medications. Even though psychologists do not prescribe medications, it is important for them to know what the various medications are, how they are used, what precautions should be taken in giving medications, and how medications affect behavior.

The elderly represent only 11% of our population; however, they account for over 20% of the estimated $10 billion spent per year on drugs and drug sundries in the United States (Fuchs, 1974; Butler, 1975). Therefore, the purpose of this chapter is to review the current knowledge of psychotropic drug use in the elderly. Its focus is largely on clinical application, rather than on basic issues in research, since formal graduate school education in psychology rarely, if ever, includes training in psychopharmacology. This omission is bewildering in our drug-oriented society and given the multidisciplinary nature of many treatment settings in

[1]The author wishes to acknowledge with great appreciation the helpful comments and suggestions of Arthur Shinn, Pharm. D., Director, Drug Information Center, St. Louis College of Pharmacy.

which clinical psychologists are employed. In addition to the
content of this chapter, the clinical geropsychologist may find
several references quite beneficial: The Pharmacological Basis
of Therapeutics (Goodman and Gilman, 1975), Clinical Use of
Psychotherapeutic Drugs (Hollister, 1973), Facts and Comparisons
(Kastrup, 1977), and Drug Treatment (Avery, 1976).

DRUG CLASSIFICATION

In order to understand the use of drugs in the elderly it
may be helpful to first review the general classification schemes
for these medications. Balter and Levine (1969) classify psychotropic drugs primarily on the basis of clinical use and secondar-
ily with respect to their chemical structure. They stress six
actions: major tranquilization, minor tranquilization, antidepressant, stimulant, sedative, and hypnotic.

The class of drugs used primarily in the treatment of psy-
choses has been called traditionally the major tranquilizers.
These drugs reduce hallucinations, delusions, and disordered
thought. The most common side effects are drowsiness, sedation,
parkinson-like syndromes (characterized by rigidity, tremor, and
weakness, and similar to the symptoms of Parkinson's disease),
and hypotension (Hollister, 1973). Major tranquilizers are not
psychologically addictive. They are rarely abused but may be mis-
used if the patient does not follow the prescribing directions of
the physician. The term ataractic has been used in the past to
describe this class of drug. Since sedation or "tranquilization"
is only one effect of this class of drugs and is not an effect of
all drugs in the class, other names have been suggested. These
include antipsychotic and neuroleptic. Currently the term neuroleptic is gaining general acceptance in the United States and is
the most accepted term internationally (Klerman, 1974). Examples
of the neuroleptic class of psychotropic drugs include the phenothiazine derivatives chlorpromazine and thioridazine, thioxanthenes
such as thiothixene, butyrophenones such as haloperidol, and the
most recent group, indolic derivatives such as molindone and
dibenzoxazepines such as loxapine succinate.

Minor tranquilizers or anxiolytics relieve anxiety without
producing excessive drowsiness. More prescriptions are filled
for minor tranquilizers than any other class of psychotropic
drugs (Ayd, 1975). These drugs can lead to physiological and
psychological dependence. Examples are the benzodiazepine derivatives, diazepam and chlordiazepoxide. The benzodiazepines
have replaced, generally, the drugs formerly used to treat
anxiety such as bromides, barbiturates, meprobamate.

Antidepressants are used to treat depression. These are
thought to be more effective against depressions which are defined
as endogenous (occurring from within). Exogenous depression (depression that is a self-limiting reaction to external events such
as loss or physical illness) on the other hand is treated more
often with minor tranquilizers rather than antidepressants. The
antidepressants do not produce a physiological or psychological
dependence. These drugs also have been called psychic energizers
or thymoleptics. Examples include imipramine, amitriptyline,
and the various monoamine oxidase inhibitors. Lithium carbonate,
although primarily used as an antimanic drug, also is thought to
have a prophylactic or protective effect against depression, especially in patients with a history of manic as well as depression.

Stimulants have been used to reduce fatigue and also to treat
depression. The uses of these drugs are largely historical; the
accepted clinical use today is in the treatment of narcolepsy
and hyperkinesis in children. Stimulant drugs are often abused
and produce a high degree of psychological dependence. The amphetamines and methylphenidate (Ritalin) are the best known examples of this class.

Sedatives and hypnotics are another class. They are used
primarily to induce sleep. In the past sedatives were used as
minor tranquilizers to control anxiety and tension. This class
of drugs has a high abuse potential. Both physical and psychological dependence can occur. The withdrawal syndrome can be
life-threatening. The use of these medications in the elderly can
produce confusion and paradoxical excitement (Eisdorfer and Fann,
1973). Examples of these drugs are chloral hydrate, methaqua-
lone, and various barbiturates. One medication, flurazepam, a
benzodiazepine, is now commonly used. It shares with other benzodiazepines a lower abuse and dependence potential.

The World Health Organization (WHO) recognizes five categories of psychotropic drugs. This classification is also clinical but claims to be based on the therapeutic mechanism of action.
The five classes are: (1) neuroleptics which have antipsychotic
action and are effective in the treatment of some psychiatric
disorders which are accompanied by neurological symptoms; (2)
anxiolytics which reduce anxiety without effecting perception or
cognition; (3) thymoleptics, used to treat depression; (4) psycho-
stimulants, used to increase levels of alertness; and (5) psyche-
delics, sometimes called psychotomimetics, which produce abnormal
mental phenomena, particularly of perception and cognition (Leavitt,
1974).

Other classifications have been suggested. Jarvik (1967) includes in his classification the categories used by Balter and

Levine but adds psychotogenic (hallucinogenic) drugs and a category for anesthetics, analgesics, and paralytics. Another classification scheme is based on the effect of psychotropic drugs on
the electroencephalograph (EEG) and is used in research (Fink,
1968). Finally, in the fields of pharmacology and medicinal chemistry, psychotropic drugs are often classified in terms of their
chemical structure.

PHYSIOLOGICAL FACTORS EFFECTING DRUG ACTION
IN THE GERIATRIC PATIENT

Various physiological factors can effect the reaction of a
geriatric patient to drug therapy. Four factors which can alter
the effects of drugs in these patients--absorption, distribution,
metabolism, and excretion--are described below.

Absorption

Absorption from the gut is a complex process influenced by
dosage form and patient parameters. For a drug to be absorbed it
first has to enter the bloodstream; that is, it must first dissolve
in the gastrointestinal tract secretion. Of the two basic kinds of
drugs (weakly acidic and basic), weakly acidic drugs such as
aspirin, barbiturates, and acetaminophen must dissolve quickly in
the acidic fluids of the stomach and duodenum for optimal absorption. One cause of decreased or delayed absorption of such weakly
acidic drugs in the elderly may be an age related reduction in the
volume of gastric acidity. This condition, known as hypochlorhy-
dria, occurs in the elderly 10 times more frequently than in younger
adults (Whitehead, 1974).

The absorption of weakly basic drugs such as amitriptyline
and diazepam depends primarily on the stomach emptying rate.
This determines how fast the drug travels from the stomach to the
small intestine. The aging process can change the number and
properties of the effector cells in the nervous system. Therefore
physiological processes such as gastric emptying, which are under
the control of the nervous system, may occur at a slower rate in
the elderly.

Due to insufficient data about drug absorption in the elderly, it is difficult to recommend specific guidelines pertaining
to age related absorption problems. However, one can assume that
elderly patients will absorb drugs less consistently or more
erratically than younger adults. Even though the rate of drug
absorption will probably be slower in the aged, the total amount

of drug absorbed will generally be the same as in younger patients. If a lessened pharmacological response or an unexpected lack of drug action is observed in an elderly patient, it may be important to consider that the absorption of the drug may be impaired due to age.

Distribution

After the drug is absorbed and enters the bloodstream, it must diffuse from the circulation to the fluids and tissues of the body and reach the site of action in order to exert a pharmacological effect. The rate of drug distribution depends on a number of factors, including the binding of the drug to protein in the plasma and the composition of the body mass.

Weakly acidic drugs such as aspirin, barbiturates, and acetaminophen are highly bound to plasma protein, especially albumin. The bound portion of the drug is pharmacologically inactive. Only the unbound portion of the drug is able to diffuse into organs and tissues or be eliminated by the liver and kidneys. In the elderly, the concentration of albumin in the plasma is decreased so that a greater fraction of the drug will be unbound in general circulation (Briant, 1977).

Drug distribution patterns may change with age because of changes in the composition of body mass. In the elderly the proportion of total body weight composed of fat increases as lean body mass, such as muscle, decreases. Certain fat soluble drugs such as phenobarbital, diazepam, and chlorpromazine are normally distributed preferentially to body fat and therefore will have a greater distribution volume in the elderly. These drugs will be stored in fat tissue to a large extent and then slowly released into general circulation. This may result in a decreased intensity and a prolonged duration of the drug action. This distribution pattern explains why geriatric patients appear to be more sensitive to the effects of these drugs when dosed according to normal adult body weight. Dosing on the basis of body weight does not take into account these age-related changes in body composition. If such highly fat soluble drugs as mentioned previously are dosed in the usual manner or are taken for prolonged periods of time, these drugs will tend to accumulate significantly in the elderly and produce toxic effects.

Metabolism

Many of the drugs given to the elderly are eliminated from the body by renal excretion. If a drug is too fat soluble to be

extracted by the kidneys, it first has to be metabolized by the
liver. Metabolism in the liver converts a fat soluble drug to a
water soluble metabolite. It has been found that the prolonged
pharmacological effects of drugs seen in geriatric patients are
usually in relation to those drugs that are metabolized by the
liver (Ritschel, 1976). The reduced rates of metabolism in the
elderly are generally attributed to reduced amounts of metaboliz-
ing liver enzymes. This could possibly arise from an inadequate
dietary intake of protein from which enzymes are synthesized by
the liver. The significance of the decreased drug metabolism in
the elderly will depend upon the nature of the drug in question.

Excretion

The rate and extent of drug excretion by the kidney depends
on three renal processes: glomerular filtration, tubular reabsorp-
tion, and tubular secretion. Of these we will discuss glomerular
filtration, the process most important to the psychopharmacology
of the elderly. Many drugs are excreted unchanged in the kidney
by glomerular filtration, a process in which a drug is passively
filtered from the blood to the renal tubules. The rate at which
the drug is eliminated is directly proportional to the rate of
glomerular filtration. This process becomes less efficient with
age. In the elderly, glomerular filtration rate has been reported
as being 25% to 50% less than in younger adults (Briant, 1977).
Thus, drugs which are mainly excreted by the kidney as active
metabolites will persist in higher concentrations and for a longer
time in the blood and tissues of the elderly patient. This declin-
ing glomerular filtration has been implicated as one of the major
factors in the toxicity seen in the elderly receiving psychotropic
drugs (Briant, 1977).

Nervous System Activity

The nervous system consists of the brain, the spinal cord,
and nerve fibers extending from both the brain and the spinal
cord. These fibers travel throughout the length of the body to
its organs such as muscles, the heart, lungs, kidneys, eyes, and
glands. The activities of all these organs are controlled by
impulses which pass along the nerve fibers. The effect of an impulse upon an organ is not direct, however. On the contrary, when
an impulse reaches the end of a nerve fiber it triggers the release of a special substance from the fiber and it is this substance which exerts an effect on the receiving organ. These substances are known as neurotransmitters. The process, however, has
three steps: (1) the neurotransmitter is released; (2) the neurotransmitter combines with the correct receptor site and stimulates

the cell; (3) the neurotransmitter is detached from the receptor site and returns to the synaptic gap and is then taken back (reabsorbed) by the nerve terminal where it originated. This re-absorption is sometimes called reuptake.

There are many neurotransmitters which play a role in the function of the nervous system. Three important substances are acetylcholine, epinephrine, and norepinephrine. The effects of the normal action of acetylcholine are called cholinergic effects. The effects of blocking the action of acetylcholine are called anticholinergic effects, resulting in dry mouth, blurred vision, and constipation. The action of acetylcholine is as follows: (1) stimulation of contractions of the intestine which push waste through the intestine for eventual evacuation; (2) stimulation of secretion of the salivary glands which release saliva into the mouth; (3) stimulation of the contraction of a muscle attached to the lens of the eye which enables a person to focus the lens on objects near to him; and (4) stimulation of contractions of the bladder which enable a person to urinate.

The name given to effects associated with epinephrine and norepinephrine are called adrenergic effects. The effects of blocking the actions of these substances are called adrenergic blocking effects. Some of the effects of epinephrine and norepi-nephrine include stimulation of the constriction of blood vessels which in turn causes the pressure of blood inside the vessels to rise, stimulation of the heart to beat faster, and stimulation of the rate of breathing.

According to Frolkis' (1977) review of aging and the auto-nomic nervous system, there is some evidence of changes in neuro-transmitters with age. These changes include: (1) a decrease in acetylcholine synthesis in postganglionic fibers; (2) highly variable or irregular changes in catecholamine metabolism such that monoamine oxidase activity is increased, catechol-o-methyl transferase remains unchanged, and catecholamine uptake by sym-pathetic nerve endings is weakened; and (3) a reduction in norepinephrine synthesis in adrenergic fibers which results in a general weakening of nervous system influences on various tissues.

CLINICAL CONDITIONS AFFECTING DRUG ACTION IN THE ELDERLY

There are a variety of patient related variables that can af-fect drug treatment in the elderly. Some of these factors are: degree of health or pathophysiological state of the individual; nature and type of medication errors; the type of medication being

used; iatrogenic factors. The problem of illness and aging is too
extensive a topic to be covered in this chapter, thus the focus
will be on medication errors and iatrogenic factors.

Medication Errors

The difficulty involved in obtaining adequate information
about the elderly patient, especially if he or she is living
alone or in an institution, increases the likelihood of misdiag-
nosis in elderly patients. Thus, medication errors (inappropriate medication ordered, dosage errors, omission, unnecessary
medication, improper timing or sequence) are much too frequent
and are rarely monitored. It is estimated that of the elderly
chronically ill attending general medical clinics, 60% are subject
to medication errors (Kapnick, 1972). More concrete examples of
medication errors include: (1) if the medication is to be taken
in the evening, the patient may be too tired to take the drug;
(2) if the medication is to be taken during a serious illness,
the patient may be too ill and weak to do so; (3) the patient may
not be able to afford the medication; (4) the patient may become
confused due to multiple medications; (5) the patient may think he
needs the medication only when he is having serious symptoms; (6)
the patient may forget to take the medication. Knowledge of these
factors will be helpful in counseling the patient about his or
her medication.

Iatrogenic (Drug-induced) Abnormalities

Psychotropic drugs in the elderly can produce a variety of
unwelcome side effects, including incontinence, oversedation, decreased motility, limb contractions, dermal necrosis, osteoporo-
sis, hypotension, peripheral neuritis, paralytic ileus, retinal
neuropathy, glaucoma, and liver degeneration. Maximum dosage
schedules should be accompanied by periodic bone marrow and liver
function tests. Disturbed elderly patients show a slow response,
or conversely an exaggerated response, when large loading doses
(large amount of drug in one dose) have been used (Berger, 1976).
The central nervous system (CNS) is the most sensitive system to
alterations in drug activity with age (Hayflick, 1976).

Alcohol and other drugs which produce primary CNS depression,
are commonly used in the elderly patient, and are associated with
acute mental symptoms include barbiturates, nonbarbiturate sedatives, hypnotics, and the nonnarcotic analgesics and the narcotic
analgesics. Dose-dependent phenothiazine-induced toxic confusional
mental states frequently have been reported (Hollister, 1973).
Confusion is generally increased if the phenothiazine is admin-

istered conjointly with other drugs possessing intrinsic anticholinergic activity.   Mental symptomatology may occur secondary to
changes in cerebral blood flow caused by orthostatic hypotension
associated with the therapeutic use of the phenothiazines, tricyclic antidepressants, or the sympatholytic antihypertensive
agents (Fann, Wheless, and Richman, 1976).   Chest infections,
often a terminal event in elderly patients, may be precipitated
because of immobility in the tranquilized, oversedated patient.
Such infections can give rise to even further confusion.   There
are some reports of confusion comparable to true toxic delirium
caused by therapeutic doses of anticholinergic drugs used to treat
parkinsonism (Gaitz, 1972).

        Insulin and the oral hypoglycemic agents (the sulfonylureas),
especially in the presence of diminished caloric intake and/or
excessive exercise, may induce marked hypoglycemia and resultant
acute mental changes.   The onset of severe hypoglycemia in the
elderly may be unheralded by the signs and symptoms of increased
sympathetic nervous system activity such as tachycardia, tachypnea,
or diaphoresis.   The patient may become comatose without warning.
Episodes of bizarre behavior, slurring of speech, disorientation,
confusion, somnulence, or inability to be easily aroused should
always be viewed with extreme suspicion in the elderly diabetic
patient receiving insulin or oral hypoglycemic agents (Hollister,
1977).   In addition, chlorpropamide has been associated with
inappropriate antidiuretic hormone (ADH) secretion which may lead
to water intoxication and give rise to changes in mental status
(Hollister, 1977).

        Prolonged thiazide therapy has been identified as a causative
factor associated with hypokalemia (potassium deficiency) and
apparent depletion of the body's potassium stores.   Potassium deficiency in the elderly can result in confusion and delirium.
Diuretics, including the thiazides, furosemide, and ethacrynic
acid, can cause marked dehydration or electrolyte disturbance
other than hypokalemia and result in impaired mental functional
capacity and acute brain syndrome in elderly patients (Briant,
1977).   Other common sources of fluid and electrolyte loss in the
elderly which may contribute to states of imbalance (dehydration)
include excessive perspiration, expectoration of copius amounts
of sputum, vomiting, diarrhea, excessive use of laxatives, inadequate fluid intake, excessive urination associated with uncon-
trolled diabetes mellitus, and hyperventilation which in some elderly patients may cause a significant loss of water via the
lungs.

        Sodium depletion secondary to diuretic therapy, especially
during periods of excessive salt restriction, may give rise to
hyponatremia (sodium deficiency).   If the serum sodium concentra-

tion is low, water shifts into the cells, the cells may become
overhydrated, their function impaired, and the patient becomes
confused (Gaitz, 1972; Pitt, 1974).

DRUG TREATMENT OF SCHIZOPHRENIA IN THE ELDERLY

Although many chronic schizophrenics live long beyond retirement age, schizophrenia usually does not have an initial onset
after 40 years of age, nor are acute episodes generally noted in
later life. Many chronic schizophrenics show at least a partial
remission of their symptoms in old age but remain hospitalized
because of the effects of prolonged institutionalization, prolonged maintenance on neuroleptic medication, and the brain changes
of old age. All three of these factors make it difficult for the
elderly schizophrenic to be resocialized into society.

Traditionally antipsychotics used to treat schizophrenia have
been divided into four major groups based upon the chemical struc-
tures of each group: (1) the rauwolfia alkaloids and derivatives;
(2) the phenothiazines; (3) the thioxanthenes and derivatives; and
(4) the butyrophenones. Two new classes of drugs now used to
treat schizophrenia are the oxoindole compounds and the dibenzoxazepine compounds.

Rauwolfia (reserpine) was the earliest antipsychotic drug
to be used but now it is more frequently used for its antihyper-
tension value. Its side effects of lowered blood pressure and
depression have led to its replacement by other antipsychotic
agents.

The prototype of most neuroleptics is the phenothiazine group
and the specific drug is chlorpromazine. Studies with the phenothiazines suggest that they act on the hypothalamus and other
subcortical centers (Black, 1970). These drugs, as is true of all
neuroleptics, have effects on the extrapyramidal nervous system.
They are believed to depress various components of the reticular
activating system and the autonomic nervous system which are involved in the control of basal metabolism, body temperature,
wakefulness, vasomotor tone, emesis (nausea) and hormonal balance
(Hollister, 1975).

Studies with the thioxanthenes suggest that they stimulate
the reticular formation, producing modifications of elicited brain
potentials and thereby shortening cortical activation. The precise mechanism of action for the butyrophenones has not been
clearly established (Hollister, 1975).

The oxoindole compounds are not structurally related to the phenothiazines, thioxanthenes, or butyrophenones. They appear to reduce spontaneous locomotion and aggressiveness, suppress conditioned responses, and serve as an antagonist to the bizarre stereotyped behavior and hyperactivity induced by amphetamines. Based on EEG studies, molindone (an oxoindole) exerts its effects on the ascending reticular activating system (Goodman and Gilman, 1975; Davis and Casper, 1977). A dibenzoxazepine compound (loxapine succinate) is also a distinct chemical structure. Behavioral studies have shown that this drug reduces aggression and provides a general calming effect (Avery, 1976).

Antipsychotics tend to act on the autonomic nervous system and produce orthostatic hypotension. They have a weak cholinergic action and it is assumed that they block access of norepinephrine and dopamine to receptors in the brain. The central nervous system effects include feelings of drowsiness and a lowering of seizure threshold. These drugs seem to be able to decrease aggressive behavior and do not tend to create psychological or physical tolerance or dependence. Large doses of antipsychotics, in contrast to sedative hypnotics, do not produce coma and anesthesia (Davis and Casper, 1977).

A variety of neurological syndromes, particularly involving the extrapyramidal system, occur following the use of almost all antipsychotic drugs. Extrapyramidal symptoms occur in 80% of patients on phenothiazines between the ages of 60 and 80 (Berger, 1976). These extrapyramidal reactions are of special concern in the elderly since they may become permanent. The extrapyramidal reactions typically respond to antiparkinson medications but sometimes concurrent treatment with phenothiazines is not recommended.

These reactions are most observable when the patient is being treated with the piperazine groups of phenothiazine drugs and with haloperidol. Acute extrapyramidal side effects occur less frequently with thioridazine and the dibenzoxazepines. Extrapyramidal symptoms may also be associated with the administration of tricyclic antidepressants.

There are four varieties of extrapyramidal syndromes associated with the use of antipsychotic medications: akathisia, akinesia, acute dystonic reactions, and tardive dyskinesia.

Akathisia is characterized by involuntary motor restlessness, constant pacing, and an inability to sit still. This reaction can be mistaken for agitation in psychotic patients; the distinction is critical since agitation can be treated with increased dosages of the medication (deGroot, 1974). Parenteral administration of benztropine allows for a differential diagnosis; agitation will not repond to this medication.

Akinesia ranges from a generalized slowing of volitional movement to a condition of immobility. The most noticeable signs are
rigidity and tremor at rest. Facial expressions are often flat and
this condition can often be misdiagnosed as depression.

Acute dystonic reactions are occasionally seen with the initi-
ation of antipsychotic drug therapy. Facial grimacing and uncoor-
dinated spasmodic movements of limbs and body can occur and may be
associated with the oculogyric crisis (an involuntary tonic contraction of the extraocular muscles characterized by a fixed upward gaze). These reactions can be misdiagnosed as either hysterical reactions of seizures, but like akathisia they respond
dramatically to parenteral administration of anticholinergic anti-
parkinsonian drugs.

Tardive dyskinesia is a late appearing neurological syndrome
associated with antipsychotic medication and occurs with an incidence as high as 20% in some patient populations (Goodman and Gilman, 1975). It is characterized by stereotyped involuntary movements consisting of sucking and smacking the lips, lateral jaw
movements, and fly catching dartings of the tongue. There also may
be purposeless quick movements of the extremities. There is no
known definitive treatment of this reaction. Although increasing
the dosage of neuroleptic may suppress the symptoms, neuroleptics
are generally discontinued as they will lead to progression of the
illness. Administration of antiparkinsonian medication should be
avoided as such treatment will exacerbate the symptoms. The use
of minimally effective dosages of antipsychotic drugs for long
term therapy, as well as discontinuation of treatment as soon as
possible, is the best preventive practice.

Antipsychotics are to be used with caution in patients who
have a history of epilepsy or ulcer disease as well as in patients
who have cardiovascular disease, respiratory impairment due to
acute pulmonary infections or chronic respiratory disorders such as
severe asthma or emphysema, or in patients with impaired liver
function. Also they are contraindicated in patients who are exposed to extreme heat or phosphorus insecticides (Hollister, 1975).

It is difficult to choose the best medication program of
antipsychotics for an elderly patient. Some suggest that the beginning dosage level be one-fourth to one-third that of the recommended adult dosage and then be increased slowly to tolerance
(Eisdorfer and Fann, 1973). In acute episodes it seems best to
begin with a low dose and increase until an effective dosage is
attained; the dosage should be reduced to the required minimum in
chronic schizophrenia. A schedule that consists of a single dose
at bedtime with drug free weekends appears to be gaining acceptance (DiMascie and Slater, 1970).

Studies of younger schizophrenics suggest that relapse rates are less if the patient is maintained on medication (Solomon and Patch, 1974). This finding is not always supported with respect to elderly patients (Eisdorfer and Fann, 1973). Distinctions between acute schizophrenia and chronic schizophrenia are important; clinical judgments must be made on the basis of individual cases. It may be that there is a group of elderly schizophrenic patients who could be described as "burned out," in the sense that the psychosis is less active. Elderly patients who have been ill for many years and who are not in a current exacerbation may not need to be maintained on any antipsychotic medication at all (Whitehead, 1974).

DRUG TREATMENT OF PARANOID STATES IN THE ELDERLY

Paranoia is often described as a feature of many disorders that occur in old age, e.g., acute and chronic brain syndromes, affective disorders. Changes in ability to process information may lead to conditions of isolation which in turn may create the appropriate conditions for paranoid reaction in the elderly. Sensory deprivation resulting from deafness and blindness may be significant contributing factors. Exhaustion due to either a debilitating illness, malnutrition, or alcoholism can further predispose the elderly patient to this type of reaction. Iatrogenic factors, particularly drug induced changes which occur in response to treatment of other conditions, can also precipitate paranoid reactions (Fann and Maddox, 1974). Examples include steroids and the drugs used in the treatment of parkinsonism, as well as the CNS stimulants used to increase alertness and motivation.

If the paranoid reaction is drug induced, withdrawal of the drug should produce a remission in the illness. If the illness is produced by sensory deprivation, amplification of sensory inputs will be highly therapeutic. If the individual is living in an isolated setting and subject to poor nutrition, a change in the environment so as to increase interpersonal relationships and provide better nutrition could easily reverse this illness. If the paranoid reaction is secondary to another illness, successful treatment of the primary illness should result in a cessation of the paranoid ideation.

The most distinctive and well defined of the paranoid states in the elderly is paraphrenia. Here the paranoia is the principal aspect of the disorder itself. It is not an exaggerated response to stress, confusion, or physical isolation. The paraphrenic patient is usually female, solitary, partially deaf and eccentric

without a past history of mental illness and usually 65 years of
age or older (Pitt, 1974). Paraphrenia is very similar to the
paranoid form of schizophrenia in which delusions are prominent.
Response to treatment with the phenothiazines, either trifluopera-
zine or fluphenazine decanoate (Prolixin), is usually good.
Other medications which have been suggested include a tricyclic
antidepressant, such as amitriptyline, or a combination of an
antipsychotic and an antidepressant, such as perphenazine and
amitriptyline (Triavil).

DRUG TREATMENT OF ANXIETY STATES IN THE ELDERLY

Anxiety in the aged may be disguised as a variety of different concerns and worries, frequently focused on somatic concerns
rather than on cognitive problems. For example, the patient may
be concerned about a failing heart, disability, or chronic pain.
In addition, concern and fear of death and dying may be another
expression of anxiety. It would appear that in the elderly anxiety
has real objects; intrapsychic conflicts may not be the cause of
the reaction.

One of the major difficulties in the treatment of anxiety reactions in the elderly is the effect of the medication used to
control the symptoms of anxiety. The drugs that are used (diazepam,
oxazepam, chlordiazepoxide, and phenobarbital or meprobamate) have
a tendency to slow the patient down and increase confusion. If
the patient's anxiety stems from having to contend with an increasingly perplexing world, further sedation may aggravate the situation and intensify the degree of anxiety. In some patients sedation may even result in aggressive behavior (Pitt, 1974).

There is a tendency for the elderly patient, once started on
an antianxiety medication, to become dependent on it. This is
especially true when barbiturates/meprobamate are used. The problems increase if the elderly person uses or abuses alcohol. The
cycle of forgetting that tablets have been taken, taking a further
dose, and then taking alcohol may increase the confusion and
anxiety, or result in acute brain syndrome or even in death. One
of the chief complications of anxiety and depression is the sleep
disturbance; such disturbances give the elderly patient a good
rationale for obtaining a hypnotic or a minor tranquilizer from
his or her physician. Caution should always be exercised in giving the elderly any type of sleep medication. In the case of
depression, the barbiturates, which are CNS depressants, may produce sleep but at the same time induce a deeper depression and
possibly accidental or intentional suicide (Eisdorfer, 1975).

Benzodiazepine Compounds

One major class of antianxiety medications includes the benzodiazepine compounds which have a depressant effect on the subcortical levels of the CNS. ALthough the exact mechanisms are not
fully understood, the calming effect of these medications suggest
activity in the limbic system (Berger, 1976). A distinctive feature of these medications is the wide margin of safety between
therapeutic and toxic doses. Ataxia and sedation occur only at
doses far beyond levels needed for antianxiety effects. Manifestations of overdose include confusion, coma, and diminished reflexes. All compounds of this group also have anticonvulsant
characteristics. Patients should be advised not to simultaneously
ingest alcohol and other CNS depressants when on benzodiazepine
medication.

These medications have not typically been used in elderly patients and there is not a great deal of information about such
usage. Although hypotension has occurred only rarely, these drugs
should be administered with caution to patients in whom a drop in
blood pressure might lead to cardiac complications. Precautions
also should be observed in patients with impaired renal or hepatic
function in order to avoid accumulation of these agents. Periodic
blood counts and liver function tests should be conducted for
patients on prolonged therapy.

In elderly or debilitated patients the initial dose should be
small and the dosage increments should be made gradually in accor-
dance with the response of the patient in order to preclude ataxia
or excessive sedation. Medications such as chlordiazepoxide are
not recommended in the same doses for the geriatric patient as
they are for other adult patients. The suggested dose for the
geriatric patient is 5 mg two to four times a day, whereas the
adult with severe anxiety or tension may take doses as high as
20 to 25 mg three to four times a day (Kastrup, 1977). The same
principles are followed for diazepam. The geriatric dose begins
at 2 to 2 1/2 mg one or two times a day and is increased very
gradually as needed and as tolerated, whereas normal adult dose
for severe anxiety and tension may range from 2 to 10 mg two to
four times a day.

Central Nervous System Depressants

Central nervous system depressants are often used to treat
the symptoms of sleep disorders. The most commonly used are the
barbiturates. Barbiturates appear to act at the level of the
thalamus where they inhibit ascending conduction in the reticular formation, thereby interfering with impulse transmission to

the cortex. Elderly patients sometimes react to barbiturates with
excitement, confusion, or depression. However, these disappear
upon withdrawal of medication. Prolonged use of these medications,
even in therapeutic doses, may result in psychological dependence.
The tendency to either misuse or abuse these substances is high in
the elderly (Benson and Brodie, 1975). Withdrawal symptoms, including delirium, convulsion, or even death, may occur after chronic
use of large doses. Therefore, these drugs should be withdrawn
gradually from any patient known to be taking excessive doses over
long periods of time. Special care is needed when any CNS depressant is given to patients who have received analgesics, other
sedatives or hypnotics, curare-like drugs, alcohol, or tranquilizers; an additive respiratory-depressant effect may ensue.

DRUG TREATMENT OF AFFECTIVE DISORDERS

The incidence of affective illnesses in the elderly is particularly high. This certainly is not unexpected in view of the
fact that depression is the most frequent response to loss; the
aged individual may suffer many losses and often has little to
compensate for them (Solomon and Patch, 1974). However, it is
important to differentiate self-limiting depressions from more
severe, longer lasting depressions characterized by prolonged and
sustained low mood and symptoms such as fatigueability, loss of
appetite, sleep disturbances, poor concentration, feelings of self-
reproach, and suicidal ideation. Self-limiting depressions, which
may be directly related to precipitating events such as the death
of a spouse or severe physical illness, may be more effectively
treated by means of psychotherapy, since they are less responsive
to somatic treatment (Jarvik, 1976). The more severe depressions,
on the other hand, may respond to somatic intervention.

A further distinction within the affective illness category
should be made between unipolar and bipolar affective states.
This distinction relates to the differentiation of manic depressive illness from pure depression. Unipolar depression refers
to patients who have had depression only. Bipolar depression
refers to patients with mania, regardless of whether they also
have had a depression (Woodruff, Goodwin, and Guze, 1974). Patients with bipolar illness have a somewhat earlier age of onset
than unipolar patients. Their histories are characterized by more
frequent and shorter episodes.

The two major classes of medications used to treat depressions
are the tricyclic compounds and the monoamine oxidase (MAO) inhibitors. Of the two classes, the tricyclics are the most widely
used in that the MAO inhibitors require dietary restrictions.

There has been some recent speculation that a third class of medication, lithium carbonate, may be useful in the treatment of depression as well as mania; however, the research in this area is
contradictory.

Tricyclic Compounds

The mechanism by which the tricyclic compounds work is not
completely known.   They are not CNS stimulants or monoamine oxidase inhibitors.   The tricyclic antidepressants inhibit the reuptake of norepinephrine into sympathetic fibers.   Since reuptake
terminates sympathetic activity, inhibition prolongs sympathetic
activity.   This action is believed to contribute to the antidepressant activity.   Although these agents are rapidly absorbed
from the gut, a clinical response may require up to three weeks.

High doses of the tricyclic compounds must be administered
with great care to patients with cardiovascular disorders.   These
medications have been reported to produce arrhythmias, sinus
tachycardia, and prolongation of the conduction time of neural
fibers (Ritschel, 1976).   Myocardial infarction and stroke have
been reported in patients taking drugs of this class (Fann, 1976).
Further, because of the atropine-like effect of these medications,
they are to be used with caution in patients with narrow angle
glaucoma which involves increased intraocular pressure.   In patients with narrow angle glaucoma even average doses of the tricyclics may precipitate an attack.

Lower than usual doses of the tricyclic compounds are recommended for elderly patients.   The dose should be gradually increased until a therapeutic level that is free of adverse side
effects is reached.   It is often appropriate to continue maintenance therapy for three months or longer to lessen the probability of relapse (Lipton, 1976).

Doxepin, previously mentioned with respect to the treatment
of anxiety, is considered primarily an antidepressant medication.
Its use is indicated in treatment of agitated depression as well as in
the treatment of depression with secondary anxiety.   Of all the
antidepressant medications, doxepin is the only one that does not
antagonize antihypertensive medications and therefore is the drug
of choice in combination chemotherapy.   At doses of up to 150 mg
per day, doxepin can be given concomitantly with guanethidine and
related compounds without blocking their antihypertensive effect,
although at dosages above 150 mg per day blocking has been reported (Fann, 1976).

The two chemical compounds that have been the model for the

development of tricyclic antidepressants are amitriptyline and
imipramine. Either of these is indicated in the treatment of unipolar endogenous depression. In adults a beginning dose of
amitriptyline is 75 to 100 mg daily with an average maintenance
dose of 150 mg daily. In general lower doses are suggested for
the elderly patient. Ten mg three times a day with 20 mg at bedtime may be satisfactory in elderly patients who do not tolerate
higher doses. An alternative would be one dose of 50 mg at bedtime. Traditionally these medications have been given in divided
doses, but recently one dose at bedtime appears to be therapeutically effective; further, it eliminates the negative consequences
of one of the drug's side effects--sedation. This dosing principle is valuable for the elderly as it may eliminate the need for
an additional hypnotic medication at bedtime. Protriptyline, a
metabolite of amitriptyline, is also available. For adults the
suggested dose is 15 to 40 mg daily. In elderly patients a 5 mg
dose three times a day is usually effective. In elderly patients
the cardiovascular system must be monitored closely if the daily
dose exceeds 20 mg.

The suggested dose of imipramine in elderly patients is about
30 to 40 mg daily with an upper limit of about 100 mg daily. (The
usual maintenance dose in younger adults is 75 to 150 mg daily.)
Desipramine, a metabolite of imipramine, is also used in the treatment of depression. Again, elderly patients can be managed on
doses smaller than those suggested for adults. The daily dose for
adults is 50 mg three times a day, up to 200 mg daily. With the
elderly, a daily dose of 25 to 50 mg, but no more than 100 mg
daily, may have the desired therapeutic effect.

Monoamine Oxidase Inhibitors (MAOI)

The monoamine oxidase inhibitors are returning to favor as a
treatment of depression. Over the last five years their use decreased significantly because of their interactions with foods and
other medications. The monoamine oxidase system is a complex
enzyme system widely distributed throughout the body which is
responsible for the metabolic decomposition of biogenic amines,
terminating their activity. Drugs which inhibit this enzyme system (MAOIs) cause an increase in the concentration of endogenous
epinephrine, norepinephrine, and serotonin in storage sites
throughout the body. It is believed that an increase in the concentration of monoamines in the CNS is the basis for the antidepressant activity of these agents. Drugs which act as MAOIs cause
a wide range of clinical effects and have the potential for serious interactions with other substances. Known hypersensitivity,
congestive heart failure, a history of liver disease or abnormal
liver function tests contraindicate the use of these medications

(Lipton, 1976).   The activation of the sympathetic nervous system
by MAOIs may result in a hypertensive crisis, one of the most dangerous side effects (Fann, 1976).   Patients taking MAOIs should
not be on multiple medications, specifically sympathomimetic drugs
including amphetamines, methyldopa, levodopa, dopamine, tryptophan,
epinephrine and norepinephrine.   Foods with a high concentration
of tryptophan (broad beans) or tyramine (aged cheese, beer, wines,
pickled herring, chicken liver, and yeast extract) should be
avoided (Kapnick, 1972).   Excessive amounts of caffeine and choco-
late can also result in a hypertensive crisis.   All patients undergoing treatment should be closely monitored for postural hypotension.   In addition to these interactions, the MAOIs also interact
with general anesthesia.   Caution is suggested with antihypertensive medication including thiazide diuretics, since hypotension
may result.   The evidence with respect to the interaction of
MAOIs with glucose metabolism or hypoglycemic agents is not clear.
Hypotensive side effects have occurred in hypertensive as well as
in hypotensive patients (Eisdorfer, 1973).

The most frequently administered drug in this category is
isocarboxazid.   The usual adult beginning dose is 30 mg daily in
divided or single dose with a subsequent reduction to 20 mg daily
for the purposes of maintenance.   A beneficial effect may not be
seen for three to four weeks.   If no effect is seen the medication
should be stopped.   Isocarboxazid also has been reported to produce
premorbid personality stabilization in the senile arteriosclerotic
patient.

Phenelzine sulfate is also employed.   The adult dose is
approximately 15 mg three times a day not to exceed 75 mg daily.
Again, three to four weeks may be necessary in order for therapeutic effects to be observed.

The third and final MAOI normally administered is tranylcypromine.   Dosage with tranylcypromine should be adjusted to the requirements of the individual patient.   Improvement should be seen
within 48 hours to three weeks after chemotherapy is initiated.   Recommended beginning doses are 20 mg per day (10 mg in the morning
and 10 mg in the afternoon).   Dosage can be increased to 30 mg
daily, 20 mg in the morning and 10 mg in the afternoon; however,
if no change is seen at this level within a week's time medication
should be stopped.

Dosage levels of these medications in elderly patients should
be determined on an individual basis.   The vast number of drug and
food interactions which may occur necessitate a valid and complete
medical and social history, including food preferences and the
presence of other medications.

Antimanic Medication

Mania (the manic phase of bipolar affective disorder) is uncommon in the elderly. Persons with bipolar affective disorder
surviving into old age will more commonly have depressive episodes rather than manic episodes. Manic episodes will occur,
however, and lithium carbonate appears to be an effective drug
for the treatment of the acute manic phase and also for the prevention of recurrent manic attacks. Lithium alters sodium trans-
port in nerve and muscle cells and effects a shift toward intra-
neuronal metabolism of catecholamines, but the specific biochemical mechanism of lithium treatment is not known at this time
(Berger, 1976). Lithium toxicity is closely related to serum
lithium levels and has a narrow therapeutic index.

The therapeutic effects of lithium carbonate generally occur
two to three weeks after onset of treatment although some changes
may be apparent around the tenth day. Antipsychotics may be helpful during this waiting period but should be discontinued after
two weeks. The therapeutic effect of lithium carbonate is a rapid
tranquilization of the person in a manic or hypermanic state.

In general many of the same pharmacologic principles which
apply to the usage of lithium carbonate in younger patients also
apply to its use with elderly patients. There are some exceptions.
The half-life of lithium in middle aged adults is about 24 hours.
In the aged it is longer--approximately 36 to 48 hours. Thus, an
elderly patient may have an adequate blood level of lithium on a
smaller oral dose than would a younger patient; hence, treatment
of the elderly patient may begin with lower doses of lithium and
levels of lithium in the blood should be carefully monitored.

The side effects of lithium are more pronounced in the geriatric patient than in the younger patient (Renschaw, 1973). Primary side effects are gastrointestinal distress (nausea) and central nervous system and neuromuscular toxicity. Confusion is a
common sign of lithium toxicity in the elderly and can occur with
relatively low blood levels of lithium. Since elderly patients excrete lithium more slowly, higher blood levels can be obtained
on a comparable dose. In adults a therapeutic blood level is between 1.0 and 1.5 mEq/L, which can be achieved with an oral dose of
1500 mg a day. In the elderly that level might be achieved with a
daily dose of 600 to 900 mg. The therapeutic range of lithium
blood levels is from 0.5 mEq/L to 1.5 mEq/L; toxicity generally
occurs at 1.5 mEq/L. However, the elderly patient may demonstrate
confusion before he reaches the top therapeutic range. Other symptoms of lithium toxicity consist of fine tremor which develops into
a coarse tremor, drowsiness, slurred speech, and confusion. In
evaluating lithium toxicity emphasis should be placed on symptoms

of lithium toxicity, rather than on blood levels. These symptoms
are extremely hard to monitor in the elderly patient living in the
community; he or she may be isolated or relatives may consider the
symptoms to be part of the "normal" aging process. In such cases
the patient may not receive the necessary intervention to prevent a
coma. No specific antidote for lithium poisoning is known. Early
symptoms of lithium toxicity can be treated by reduction or cessation of dosage. After a 24 to 48 hour wait, medication can be
resumed at lower doses.

TREATMENT OF BRAIN SYNDROMES

In terms of chemotherapy for acute and chronic brain syndromes,
only the smallest effective doses of drugs acting on the central
nervous system should be used; patients suffering from a brain syndrome are frequently sensitive to these agents (Hollister, 1977).
Brain syndromes are, in fact, often precipitated by sedatives or
hypnotics and may subside when such drugs are discontinued. No
drug is entirely safe. Careful attention to dosage and mental
status is more important than the particular sedative or hypnotic
used.

A number of vasodilators and CNS stimulants have been used to
improve learning, reduce confusion, and facilitate memory in older
patients. These substances include ergot alkaloids (hydergine),
pentylenetetrazol combined with niacin and vitamins, magnesium
pemoline, methylphenidate, deanol, and amphetamines (Hollister,
1977). Their effectiveness, however, is questionable.

The ergot alkaloids have been suggested for use in patients
who show confusion, unsociability, dizziness, and other symptoms
characteristic of arteriosclerosis or cerebral insufficiency. Most
studies have been short term, poorly controlled, and suggest modest
improvement in performance in the above areas (deGroot, 1974). The
mechanism of the drugs' actions which improve performance is not
presently understood and their use is controversial.

Pentylenetetrazol, in small doses, is a mild CNS stimulant.
Niacin (nicotinic acid) is added because of its vasodilating effect;
vitamins per se have not been shown to have specific benefits with
respect to brain syndrome symptoms, except in those involving a
known vitamin deficiency. These products are intended for use in
the treatment of a wide variety of nonspecific symptoms in the
elderly. These symptoms include senile confusion, depression, functional memory defect, antisocial attitudes, and general debilitation.
Reports of the efficiency of these products are mixed (Hollister,
1977). It would seem that there is sufficient evidence to suggest

further research in this area. Important aspects of this type of
research would involve both psychological measurement of improved
performance as well as direct measures of increased blood flow.

CNS stimulants all operate in a manner similar to amphetamines (sympathomimetic amines which stimulate activity of the
central nervous system). Peripheral actions include elevation of
systolic and diastolic blood pressures and stimulation of bron-
chodilator and respiratory action. At one time these medications
were used in the treatment of depression; however, due to their
short term therapeutic effect and their high abuse potential they
are no longer recommended. These drugs may decrease the hypotensive effects of guanethidine and interact with insulin and other
antidepressants to produce a hypertensive crisis. Confusion,
agitation, and cardiovascular arrhythmias are all signs of over-
dosage as well as adverse side effects (Ayd, 1975; Anderson and
Judge, 1974). These medications can produce sleep disturbance
and induce insomnia.

ELECTROCONVULSIVE THERAPY

Electroshock therapy (EST), now usually called electroconvulsive therapy (ECT), is the second most widely used form of
physical therapy. Many different techniques of producing convul-
sions, primarily chemical, were used prior to 1938 when two
Italian psychiatrists found that passing an alternating electri-
cal current through the head and brain was the most reliable, con-
venient, and effective method for producing a convulsion in a
patient. In the United States ECT was first used in 1940 (Anderson and Judge, 1974).

The standard procedure consists of the passage of small
measured amounts of electricity through two electrodes placed on
the temporal areas of the patient's head. The patient is anes-
thetized with a short acting barbiturate (e.g., hexabarbital) and
his muscular activity is decreased by the use of a neuromuscular
blocking agent, succinylcholine chloride. The voltage varies
from 70 to 130 volts, duration from 0.5 to 2.0 sec, and the amount
of current from 200 to 1600 milliamperes. The procedure induces
in the patient a brief tonic phase (10 sec) followed by a clonic
phase (30 to 40 sec). The optimal dosage is the lowest current for
the shortest time that will produce a grand mal or generalized
tonic and clonic convulsion. Typically three treatments a week
are given until substantial improvement occurs and then another
three or four treatments complete the course. A minimum of six
treatments, an average of nine treatments, and a maximum of 25
treatments are considered normal limits for a course of treatment
(Solomon and Patch, 1974).

The mechanism of action of ECT is not known; it is an empirical treatment modality that appears to be helpful in terminating certain types of depressive episodes. Seizure activity seems to be necessary for beneficial effects to be realized. According to Kral (1976), the number of ECT treatments should be kept to a minimum and treatment should be terminated at the first signs of improvement; further treatment can be conducted by means of antidepressant medications.

ECT is used primarily as a treatment for depression, although some psychiatrists use ECT to treat acute schizophrenic episodes, an excited stage of a manic reaction, or a catatonic reaction. Some clinicians suggest that ECT is superior to antidepressant medication in that it provides a more rapid response and a greater percentage of responses in depressed patients (Solomon and Patch, 1974; Avery and Winokur, 1976).

According to Solomon and Patch (1974), the only absolute contraindication to ECT is brain tumor; the increased intracranial pressure that occurs during the convulsion could have dire consequences. In diseases in which there is question about treatment with ECT one must weigh the possibility of death from exhaustion, suicide, or starvation against the danger of therapy.

Electroconvulsive therapy should be administered with caution to patients with bone and joint disease; however, the risks involved in such patients have been minimized in the treatment as currently given, using a neuromuscular blocking agent as a premedication. In cases where there is known cardiac decomposition such as myocardial disease, an abnormal EKG, angina, or coronary thrombosis, the heart disease must be treated before ECT is undertaken. Caution is advised in using ECT with patients suffering from recent myocardial infarction, severe congestive heart failure, aortic aneurysm, complete heart block, or ventricular tachycardia. Atropine is generally given pretreatment to decrease cardiac irritability.

Fear of treatment as well as post-treatment confusion and memory loss can further complicate the use of ECT with elderly patients. Fear of treatment may generalize to noncompliance with all types of therapy, thus creating a more serious management problem than existed before treatment. Confusion and memory loss may also create more severe management problems. Confusion generally occurs after each treatment but tends to clear. When treatments are spaced at three per week memory loss persists in the interval between treatments after about the sixth treatment. The nature of the memory loss is retrograde, and appears to affect recent or short term memory more than long term memory (Abrams, 1972). It is suggested that memory loss clears in about three weeks after treatment without permanent damage. However, Friedberg (1976)

cautions that ECT may, in fact, cause brain damage and that memory
function may not return to pretreatment levels. This potential
side effect should be given serious consideration prior to the
administration of ECT to aged patients, many of whom may experience
difficulty with short term memory as a function of "normal" aging
or in response to other disease processes (Kral, 1976). In recent
years a number of studies have indicated that unilateral treatment
on the nondominant side produces a bilateral seizure which is
associated with less memory loss and less confusion (Abrams, 1973;
Fleminger, 1970). However, more treatments generally are required.

CONCLUSIONS

Extra caution should be used when treating the elderly with
psychotropic medications. The efficiency of the body and individual
organs lessens with age; consequently, the ability of the elderly
patient to absorb, detoxify, and excrete drugs is reduced. Excretion and detoxification in particular are directly affected with
increased age. The basis of altered reactions to drugs in the elderly are varied: (1) individual pharmacokinetics show greater
variability due to reduced renal and hepatic function and altered
enzyme function; (2) the number of interaction drugs is significantly increased with age; (3) cerebral blood flow is significantly
reduced in the cerebral arteriosclerotic patient; and (4) there is
a loss of neurons in the brain which may interfere with the interactions and balance of the various cerebral centers and there produce paradoxical or unusual reactions to psychotropic medications.

The use of drugs in the elderly patient should follow some
simple rules:

1. Know the pharmacological action of the drug being used and
    particularly how it is metabolized and excreted.

2. Use the lowest dose that is effective in the individual pa-
    tient; drug dose should be titrated with patient response.

3. Use the fewest drugs the patient needs.

4. Do not use drugs to treat symptoms without first discovering
    the cause of the symptoms; sometimes reactions may be due to
    a social deprivation which cannot be corrected by medication.

5. Do not withhold medication because of age per se, particularly
    when drug therapy may improve the person's quality of life.

6. Do not use drugs if the symptoms caused by the drug are worse
    than those the drug is suppose to relieve.

7. Do not continue to use a drug if it is no longer necessary.

8. Review repeat prescriptions at least quarterly in elderly
    patients.

9. Encourage the patient to call if he has an adverse side ef-
    fect or if he does not remember how or when to take the medi-
    cation. Be sure the prescribing physician is willing and
    available to listen to the patient's questions when he or
    she does call.

Table 1.   Generic Drugs With Their Most Commonly Used Trade Names

Generic	Trade	Generic	Trade
MAJOR TRANQUILIZERS		**ANTIDEPRESSANTS** (cont'd.)	
		MAO Inhibitors	
Chlorpromazine	Thorazine		
Fluphenazine	Prolixin		
Haloperidol	Haldol	Isocarboxazid	Marplan
Loxapine	Loxitane	Phenelzine	Nardil
Mesoridazine	Serentil	Tranylcypromine	Parnate
Molindone	Moban		
Perphenazine	Trilafon	**ANTIMANICS**	
Prochlorperazine	Compazine		
Thioridazine	Mellaril	Lithium Carbonate	Eskalith
Thiothixene	Navane		
Trifluoperazine	Stelazine	**STIMULANTS**	
		Amphetamine	Benzedrine
MINOR TRANQUILIZERS		Dextroamphetamine	Dexedrine
		Methylphenidate	Ritalin
Benzodiazepines (Antianxiety)			
		SEDATIVES AND HYPNOTICS	
Chlordiazepoxide	Librium		
Clorazepate	Tranxene	_Nonbarbiturate_	
Diazepam	Valium		
Oxazepam	Serax	Chloral Hydrate	Noctec
		Ethchlorvynol	Placidyl
Mephesine-like Compounds		Flurazepam	Dalmane
		Gluthethimide	Doriden
Meprobamate	Equanil	Methaqualone	Quaalude
	Miltown		Sopor
		Methyprylon	Noludar
Sedating Antihistamines			
		Barbiturate	
Hydroxyzine	Atarax		
	Vistaril	Amobarbital	Amytal
Promethazine	Phenergan	Amobarbital and	Tuinal
		Secobarbital	
ANTIDEPRESSANTS		Pentobarbital	Nembutal
		Phenobarbital	Luminal
Tricyclics		Secobarbital	Seconal
Amitriptyline	Elavil	**ANTI-PARKINSONISM AGENTS**	
Desipramine	Norpramin		
Doxepin	Adapin	Amantadine	Symmetrel
	Sinequan	Benztropine	Cogentin
Imipramine	Tofranil	Carbidopa and	Sinemet
Nortriptyline	Aventyl	Levodopa	
Perphenazine and	Triavil	Levodopa	Dopar
Amitriptyline	Etrafon		Larodopa
Protriptyline	Vivactil	Trihexyphenidyl	Artane

REFERENCES

Abrams, R. Recent clinical studies of ECT. *Seminars in Psychi-
    atry*, 1972, 4, 3-12.
Anderson, W. F. and Judge, T. G. *Geriatric medicine*. New York:
    Academic Press, 1974.
Avery, D. *Drug treatment*. Sydney, Australia: Adia Press (in
    collaboration with Acton, Mass.: Sciences Group), 1976.
Avery, D. and Winokur, G. Mortality in depressed patients treated
    with electroconvulsive therapy and antidepressants. *Archives
    of General Psychiatry*, 1976, 33, 1029-1037.
Ayd, F. J. *Rationale psychopharmacotherapy and the right to
    treatment*. Baltimore: Ayd Medical Communications, Ltd.,
    1975.
Balter, M. and Levine, J. The nature and extent of psychotropic
    drug usage in the United States. *Psychopharmacology Bulletin*,
    1969, 5, 3-13.
Benson, R. A. and Brodie, D. C. Suicide by overdoses of medicines
    among the aged. *Journal of the American Geriatrics Society*,
    1975, 23, 304-308.
Berger, F. M. Present status of clinical psychopharmacology.
    Clinical Pharmacy and Therapeutics, 1976, 19, 725-731.
Black, P. *Drugs and the brain*. Baltimore: John Hopkins Press,
    1970.
Briant, R. H. Drug treatment in the elderly: Problems and pre-
    scribing rules. *Drugs*, 1977, 13, 225-229.
Butler, R. N. Psychiatry and the elderly: An overview. *Ameri-
    can Journal of Psychiatry*, 1975, 132, 893-900.
Davis, J. M. and Casper, R. Antipsychotic drugs: Clinical phar-
    macology and therapeutic use. *Drugs*, 1977, 14, 260-282.
de Groot, M. H. L. The clinical use of psychotherapeutic drugs in
    the elderly. *Drugs*, 1974, 8, 132-138.
DiMascio, A. and Shader, R. I. (Eds.). *Clinical handbook of psycho-
    pharmacology*. New York: Science House, 1970.
Eisdorfer, C. Observations on the psychopharmacology of the aged.
    Journal of the American Geriatrics Society, 1975, 23, 53-57.
Eisdorfer, C. and Fann, W. E. (Eds.). *Psychopharmacology and
    aging*. New York: Plenum, 1973.
Fann, W. E. Pharmacotherapy in older depressed patients. *Journal
    of Gerontology*, 1976, 31, 304-310.
Fann, W. E. and Maddox, G. L. *Drug issues in geropsychiatry*.
    Baltimore: Williams & Wilkins, 1974.
Fann, W. E., Wheless, J. C., and Richman, B. W. Treating the aged
    with psychotropic drugs. *Gerontologist*, 1976, 16, 322-328.
Fink, M. EEG classification of psychoactive compounds in man: A
    review and theory of behavioral associates. In D. Efron
    (Ed.), *Psychopharmacology*, U.S. PHS Pub. No. 1836, 1968.
Fleminger, J. J. Differential effect of unilateral and bilateral

ECT. American Journal of Psychiatry, 1970, 127, 437–442.

Friedberg, J. Shock treatment is not good for your brain. San Francisco: Glide Publications, 1976.

Frolkis, V. V. Aging of the autonomic nervous system. In J. Birren and K. W. Schaie (Eds.), The handbook of the psychology of aging. New York: Van Nostrand Reinhold, 1977.

Fuchs, V. Who shall live. New York: Basic Books, 1974.

Gaitz, C. M. Aging and the brain. New York: Plenum Press, 1972.

Goodman, L. S. and Gilman, A. The pharmacological basis of therapeutics. New York: MacMillan, 1975.

Hayflick, L. The cell biology of human aging. The New England Journal of Medicine, 1976, 295, 1302–1308.

Hollister, L. E. Clinical use of psychotherapeutic drugs. Springfield, Ill.: Charles C Thomas, 1973.

Hollister, L. E. Drugs for mental disorders of old age. Journal of the American Medical Association, 1975, 234, 195–198.

Hollister, L. E. Mental disorders in the elderly. Drug Therapy, 1977, 7, 128–135.

Jarvik, L. F. Aging and depression: Some unanswered questions. Journal of Gerontology, 1976, 31, 324–326.

Jarvik, M. The psychopharmacological revolution. Psychology Today, 1967, 1, 51–59.

Kapnick, P. L. Age differences in opinions of health care services. The Gerontologist, 1972, 12, 294–297.

Kastrup, E. K. (Ed.). Facts and comparisons. St. Louis: Facts and Comparisons, Inc., 1977.

Klerman, G. L. Neuroleptics: Too many or too few. In F. Y. Ayd, Jr. (Ed.), Rational psychopharmacology and the right to treatment. Baltimore: Ayd Medical Communications, Ltd., 1975.

Kral, V. A. Somatic therapies in older depressed patients. Journal of Gerontology, 1976, 31, 311–313.

Leavitt, F. Drugs and behavior. Philadelphia: W. B. Saunders, 1974.

Lipton, M. A. Age differences in depression: Biochemical aspects. Journal of Gerontology, 1976, 31, 293–299.

Pitt, B. Psychogeriatrics: An introduction to the psychiatry of old age. London: Churchill Livingstone, 1974.

Renschaw, D. C. Depression in the 1970's. Diseases of the Nervous System, 1973, 34, 241–245.

Ritschel, W. A. Pharmacokinetic approach to drug dosing in the aged. Journal of the American Geriatrics Society, 1976, 24, 344–354.

Solomon, P. and Patch, V. D. Handbook of psychiatry. Los Altos: Lange Medical Publications, 1974.

Whitehead, J. M. Psychiatric disorders in old age. New York: Springer Publishing Company, Inc., 1974.

Woodruff, R. A., Goodwin, D. W., and Guze, S. D. Psychiatric diagnosis. New York: Oxford, 1974.

BEHAVIORAL APPROACHES TO THE PROBLEMS OF LATER LIFE

William S. Richards and Geoffrey L. Thorpe

The Counseling Center

Bangor, Maine

The unifying feature of the various procedures described by
"the behavioral approach," "behavior therapy," "behavior modification," "applied behavior analysis," and "social learning theory,"
is the general approach taken to the assessment and clinical management of undesired, intolerable, or dysfunctional behavior. In
this context, the term "behavior" is used to underscore the
fact that, ultimately, problems are manifested in some area of
activity, whether verbal or non-verbal; the formulation does not
preclude examination and treatment of problems in emotion,
cognition, or imagery. Originally, "behavior therapy" referred
to the practice of specific therapeutic procedures which were
developed in a two-stage process: First, clinical applications
were sought of certain principles explicated by experimental
psychologists (particularly the principles of learning theory);
and second, the resultant procedures were empirically evaluated
in their own right in clinical--and in analogues of clinical--
settings (cf. Eysenck and Rachman, 1965; Wolpe and Lazarus,
1966). Recent formulations of behavior therapy have explicitly
encompassed a broader range of human functioning (e.g. "multimodal behavior therapy," "cognitive-behavior modification,"
and "biofeedback"), and the term now refers more to an approach
to treatment than to a finite list of techniques. The field has
been extensively described in texts (e.g. Bandura, 1969; Franks,
1969; O'Leary and Wilson, 1975; Rimm and Masters, 1974), in
journals devoted to the field (e.g. Behavior Therapy; Behaviour
Research and Therapy; Journal of Applied Behavior Analysis;
Journal of Behavior Therapy and Experimental Psychiatry), and
in at least one continuing annual review of the area (Franks and
Wilson, 1973, 1974, 1975, 1976). A salient aspect of the behavioral approach is the emphasis on specificity: "How may which

253

behaviors in <u>what</u> individual be most effectively and efficiently
altered by <u>whom?</u>" summarizes the typical questions. It remains to
be seen how seriously behavioral researchers have taken these concerns in application to the problems of later life.

It seems to be true of the elderly as a group that, although
the need for mental health services is especially evident in later
life, the availability--or at least the utilization--of such services is less than for any other age group (Butler and Lewis, 1977;
Gottesman, 1977; Group for the Advancement of Psychiatry, 1971;
Lawton and Gottesman, 1974). It is certainly true that, as a sub-
specialization, behavior therapy has not distinguished itself in
the volume of work produced relating to problems in the elderly.
Reasons for this neglect lie at least partly in the way behavior
therapists conceptualize their patients: (1) traditional psychiatric diagnosis provides no prescription, in itself, for treatment
by behavior therapy, although for other professionals, the elderly
may be distinguished as a group by the kinds of diagnosis they often
receive (Roth, 1976); (2) an examination of the literature of behavior therapy shows relatively little preoccupation with either
chronological age or "psychosocial stages" as prescriptive variables: A technique such as assertion training, for example, is "a
problem-solving therapy, not a population-oriented one" (Corby,
1975); and (3) in view of the above, "over 60" or "over 65," to the
behavior therapist, denote an arbitrary division of the human race.

Nevertheless, sound reasons for the particular relevance of
behavioral approaches to the problems of later life have been provided. On the negative side, long-term "depth" psychotherapy is
impractical, and the high base-rate of syndromes implying an "organic"
etiology in the elderly is discouraging to many clinicians who en-
dorse formal medical diagnosis as prescriptive of treatment, whereas
behavior therapy is contraindicated by neither concern (Cautela and
Mansfield, 1977). More positively, behavior therapy emphasizes
therapist activity, preferred in general by relatively less-active
older people; the stresses of later life often produce anxiety,
which is particularly well managed by behavioral approaches; and
concepts of "self-control," emphasized by behavioral practitioners,
seem particularly relevant to the extent that older people may be
less effective in controlling their environments (Cautela and Mansfield, 1977; Riedel, 1974). Several other writers have argued in
favor of the application of behavior therapy to the problems of later
life (Barnes, Sack, and Shore, 1973; Cautela, 1966, 1969; Cohen,
1967; Hoyer, 1973; Hoyer, Mishara, and Riedel, 1975; Lindsley, 1964),
although it remains to be seen to what extent these exhortations
have been heeded.

Promising areas for therapeutic intervention in the elderly
appear to be (1) those problems which may be experienced at any time
of adult life (e.g. marital problems, anxiety states, addictive

behaviors); (2) problems related to loss or bereavement (e.g. depression, grief, role-performance dysfunction); (3) problems associated with brain dysfunction (e.g. aphasia, memory loss); (4) coping with chronic conditions (e.g. cardiovascular dysfunction, hearing loss); and (5) multiple problems (see Roth, 1976, for a general overview of the relevant psychiatric disorders).

In practice, much of the literature to be covered in this chapter reports on treatment of institutionalized groups. This raises the concern that, because any complained-of behavior may be evaluated for behavioral treatment, behaviors seen as problems by people other than the patient (e.g. relatives, nursing-home staff, hospital staff) may be selected for intervention over and above those presented by the elderly clients themselves. Thus, "socially-significant" behaviors (e.g. incontinence, demanding or manipulative behavior) may be targeted more often than "clinically-significant" behaviors (e.g. intrusive imagery, depressive cognitions). This risk may be at its greatest in the case of an older person, possibly with multiple problems and a history of long-term hospitalization. Whereas in the typical institutional milieu the staff may be relatively less sensitive to the issue of "informed consent," it is gratifying to note that behavior therapists, with their experience of such potential hazards in the context of the token economy, for example, have addressed issues of potential abuse of this kind (Franks and Wilson, 1975, Section I; 1976, Section I). Some have attempted to specify general characteristics of the behavioral approach which militate against unethical or coercive behavior on the part of the therapist. These include: The goal of the alleviation of human suffering and the enhancement of human potential; the effort to facilitate improved functioning as indicated by increased skill, independence and satisfaction; and the practice of being guided by a contract between client and therapist in which the goals and techniques of therapy are agreed (Davison and Stuart, 1974). It will be well to remember these points in evaluating the material reviewed here.

Other writers, with varying perspectives and emphases, have attempted to systematize the literature on behavior therapy with the elderly. Cautela and Mansfield (1977) give recommendations and suggestions for the use of behavioral procedures typically employed in individual therapy. Based on their own case studies, they give examples of the application of systematic desensitization and covert therapy procedures with elderly individuals who might be seen in an outpatient setting, or who might be referred for individual treatment by nursing-home staff. Their review is restricted to treatments based on respondent and covert conditioning. Riedel (1974), on the other hand, provides a detailed review of operant learning procedures as applied to middle-aged and older populations, with emphasis on problems often presented in institutional settings. These include dysfunctional social behavior in the ward

community, work performance, self-care skills, etc., as opposed to
the affective or cognitive disturbances that might be presented in
an outpatient context (i.e. the kind of problem addressed by
Cautela and Mansfield). The work reviewed by Riedel centers around
a "rehabilitation" model. Finally, Lindsley (1964), rather than
attempting to change the elderly individual, helping him/her to
accommodate to a progressively less-manageable environment, suggests the design of a "prosthetic environment" in which important
features of the physical environment are arranged in a way which
promotes and supports functional behavior. Thus, losses in sensa-
tion, memory, attention, and visuomotor coordination are rendered
less handicapping by the provision of multiple sense-modality dis-
plays, fail-safe devices in the occupational therapy setting,
response-force amplifiers, and so forth.

     In summary, behavior therapy is regarded as an approach to
treatment encompassing a variety of techniques and strategies; our
definition of this field includes, but is not limited to, the above
conceptualizations. Rather, we would wish to emphasize the fact
that, in general, behavioral approaches are problem-oriented rather
than population-oriented; and that the use of the term "problem"
does not preclude attempts to foster positive "growth" and self-
improvement in a population which has, unfortunately, been all too
often regarded in a pessimistic light. Further, we would hope to
model after Riedel (1974), who has attempted to interweave a
growth-oriented behavioral approach with the optimistic concept of
"life-span development." This view stands in contrast to the
"debilitation" or "decline" model of aging.

     This chapter attempts a brief overview of behavior therapy for
the elderly in each of four areas of functioning: Social behavior;
physical problems and the sensory-motor domain; hygiene, self-care,
and self-management; and cognitive functioning and emotionality.
A final section reviews studies of the modification of staff behavior and attitudes.

                          SOCIAL BEHAVIOR

Group Behavior Therapy and the Token Economy

     Whether or not the client's natural setting is within a de-
fined group (which may or may not include other elderly people),
there are clear indications for treating the problem "social" behaviors of the elderly in group contexts. The fact of age-related
communication changes (for the worse) seems well-established, and
group therapy has been recommended as a particularly effective
remedial vehicle (Bollinger, 1974). In a study of aged people
without manifest problems, it was found that "primary group inter-

action" (i.e. the extent of social contact with relatives, friends, or neighbors) contributed to their general involvement in life activities (Hampe and Blevins, 1975). In a two-year study of institutionalized aged, it was found (among other results) that patients who did not decline in general intellectual skills were more socially reactive at the start of the observation period (Kleban, Lawton, Brody, and Moss, 1976). Although this correlation implies no causality, let alone the direction of causality, this relationship at least suggests that maintaining social responsiveness may bring general benefit. A multi-faceted program involving a variety of social encounters (including sexual relations even for institutionalized patients) has been commended as a deterrent to suicide in the aged (Leviton, 1973). In a masterly and influential review of "the chronic mental patient"--a group that includes many elderly people--Paul (1969) concludes that "milieu" and "social learning" approaches will make for the best general prognosis.

Unfortunately, comparatively few reports are available of research on behavior therapy in groups, and this is particularly true in the context of the problems of later life. Franks and Wilson (1973) noted a dearth of satisfactory papers on group treatment in the entire behavior therapy arena, despite the fact that

"It is our conviction not only that behavioral methods lend
themselves particularly well to group implementation but
that such a strategy could well result in uniquely potent
behavior change" (Franks and Wilson, 1973, p. 481).
To this we might add: ". . . and particularly so for an elderly population."

It is also important to acknowledge a sobering caveat in evaluating at least some of the studies of behavioral intervention which are available: The phenomenon of "trivialization," in which methodologically impressive studies of behavior therapy for the aged are, in reality, of limited relevance or significance (Kahn, 1977). Kahn refers to the ubiquitous "Hawthorne effect," and is eloquently derisive of one study in which the researchers found that the availability of a heavily-advertised ward store was, not surprisingly, followed by an increase in the number of patients to be found in the vicinity of the store when it was open. Kahn regards this kind of trivialization as a sad indictment of both the level of treatment available for the elderly in institutions, and the level of expectation researchers and clinicians have for this group of patients. A number of studies of this kind were discarded from our review material.

Two relevant papers from 1972 are strictly applicable to "chronic" psychiatric patients, although some patients in each sample were over 65. Olsen and Greenberg (1972) found that group treatment with "incentives" and reinforcement produced superior performance to groups with "interaction" only, and to a "milieu"

control group, at least on the behavioral criteria of number of
off-grounds passes taken by the hospitalized patients and their
participation in "work details," whereas, on a possibly more interesting but less reliable "social adjustment" scale, the "interaction" group was more successful. A study by Weinman, Gelbart,
Wallace, and Post (1972) is particularly interesting because it
produced a differential response to treatment on the part of the
older patients. There were three treatment conditions, each administered for three months: A "socioenvironmental" treatment, in
which patients took part in weekly group activities, stimulated
and encouraged by ward staff; a "desensitization" condition, in
which assertive behaviors formed the basis of an anxiety-hierarchy
which was presented imaginally during relaxation; and a "relaxation" condition, in which 36 sessions of relaxation training were
conducted. Results indicated that the socioenvironmental approach
was most effective for inducing assertive social behavior, particularly for the older patients. This study incidentally raises
another issue: It illustrates quite clearly that for the population under investigation, many more sessions of treatment are
needed than for a younger group; the 36 desensitization sessions
were insufficient for the completion of the 10-item anxiety-hier-
archies in the desensitization group. Similar findings of important differences between the response to treatment of old versus
young patients are reported by Cautela and Mansfield (1977), from
case-report data. A related caution is raised by Hutchison (1974),
who found that a priori assumptions that behavior in the elderly
may be strengthened by "social reinforcement" (praise or reproof)
were disconfirmed. To counteract some of the problems of working
with this specific population, Cautela and Mansfield provide some
thoughtful ideas about relevant kinds of items to consider in
desensitization hierarchies and in covert therapy procedures, and
Corby (1975) does the same for situations calling for interpersonal
assertiveness. In any case, the Weinman et al study does substan-
tiate the claim that active, group methods are appropriate for
addressing problems of interpersonal expressiveness in the elderly.

     In a later study, Hoyer, Kafer, Simpson, and Hoyer (1974),
working specifically with elderly mental patients, successfully
reinstated speech by reinforcing verbalizations in a small-group
context. Positive generalization to the ward context was found.

     The token economy, a more formal and structured way of speci-
fying the contingencies for a ward-group, seems potentially as
applicable to the elderly as to younger patients, although there
are concerns about the meaningfulness of "informed consent" to participation in such a program in at least some older patients (see
Introduction, above). Most reported token economies do not differentiate their subject populations into older and younger groups,
because, similarly to "assertion training" as noted above, the
token economy is oriented towards specific problems rather than to

particular diagnostic or age groups. In fact, token economies have
been applied to practically all conceivable patient-groups (Franks
and Wilson, 1976, Section VII). Essentially favorable results--at
least in terms of increasing social participation, although "symp-
tomatic" behaviors remained unchanged--were obtained in a typical
token economy in which two of the seven experimental patients were
over 60 (Fraser, McLeod, Begg, Hawthorne, and Davis, 1976). The
same kind of result, from a study of patients ranging in age from
39 to 72, was obtained by Mumford, Patch, Andrews, and Wyner
(1975), who also found improvements in social functioning without
detecting changes in specific behaviors associated with psychopathology.

Franks and Wilson (1976, Section VII) review two studies in
which it was shown that the provision of recreational materials,
and staff prompting of nursing-home residents to use them, was
effective in stimulating this kind of social participation. They
note that merely providing the materials was ineffective, whereas
there is evidence that this would not be the case for a younger
population. It is quite possible, in view of this, that the subjects in the Weinman et al (1972) study benefited especially from
the staff's prompting and encouragement of social responsiveness
in their "socioenvironmental" therapy.

Individual Treatment Methods

Behavior therapy procedures commonly used in out-patient
clinical practice include the systematic desensitization of situational anxieties, modeling and behavior rehearsal of effective
interpersonal behavior, and the "cognitive restructuring"
(Lazarus, 1971) of handicapping, defeating cognitions about one's
personal worth, often implicated in depression or in deficient
social functioning. Cautela and Mansfield (1977) have provided
examples of applications of such procedures to social problems in
later life, showing that the techniques are suitable for presentation to elderly patients, provided that appropriate modifications
of procedure are made. Relaxation training is recommended specifically for geriatric patients as a procedure which brings general benefit (Richard, Picot, deBus, Andreoli, and Dalakaki,
1975), but these authors presented no precise data on its effectiveness. Techniques based on operant conditioning have received
most attention for application to dysfunctional social behavior
in later life, although in a major review of this specific area
covering the literature through 1973, Riedel (1974) found only a
handful of studies. Three such studies demonstrated the efficacy
of reinforcement procedures for reinstating speech in older
psychiatric patients in institutions (Baker, 1971; Hoyer et al,
1974; Sabatasso and Jacobson, 1970). The Hoyer et al study, as
noted earlier, was conducted with small groups of patients.

Similar procedures were used to treat verbally over-productive behavior, chronic screaming, in an 80-year-old nursing-home resident (Baltes and Lascomb, 1975). Whenever no screaming was heard in a two-minute period, nursing staff praised, touched, and smiled at the patient, or gave her attractive food and played music on a tape-recorder. Whenever the patient screamed, the nurses turned their backs on the patient, and put the reinforcers out of sight. By comparing the rate of screaming with and without the treatment contingencies, the authors demonstrated clearly that the procedures used were effective in reducing the undesired behavior.

Finally, some authors argue in favor of a multi-faceted behavioral approach to disturbed aged individuals without, however, providing experimental data. Preston (1973) gives an overview of some relevant behavior therapy procedures, and Brody, Cole, and Moss (1973), without subscribing to an avowedly "behavioral" orientation, stress the merits of a thorough behavioral assessment and a multimodal approach to redressing the deficits found in each area of functioning. The main purpose of the paper is to discourage therapeutic defeatism when confronted by mentally impaired elderly patients.

PHYSICAL PROBLEMS AND THE SENSORY-MOTOR DOMAIN

Even within this medical/physiological domain, behavior therapy approaches have been shown to have some applicability (Peck, 1972, 1976). Dependent or "patient" behaviors commonly develop to a greater extent than is warranted among individuals who acquire physical problems in later life; the presumed etiology of such behaviors is that they become controlled by their consequences, in the form of increased attention from the care-giving staff. In addition, physical problems often lead to the generation of concomitant emotional problems (e.g. a sense of loss of personal worth; subjective discomfort about the untoward stares of others). Loss of physical function in one area often requires the development of alternative skills, using other sensory or motor modalities. Each of these concomitants of physical disability is amenable to modification, at least in principle, by the use of behavioral techniques.

The older client is more likely to have medical/physical problems, and to have disproportionately more chronic problems, than younger patients. Only recently has psychological treatment been regarded as appropriate in areas usually considered to fall within the medical domain. Paradoxically, although older clients are assailed by more frequent, more debilitating, more complex, and more chronic physical problems, behavioral techniques have been more commonly applied to the physical problems of younger individuals.

The problem of learned dependency, or "learned helplessness" (Seligman, 1975), is often of primary concern to care-giving staff. The hospital or institutional environment provides, often enough, the very contingencies that foster learned helplessness. Ironically, ward staff tend to be particularly concerned about this syndrome which is generated, in turn, by their own behavior, as well as other features of the institutional setting.

Three relevant reports deal with low rates of ambulation in certain nursing-home residents or hospitalized psychiatric patients (Burt, Law, Machan, Macklin, Nesbitt, Read, and Wiebe, 1974; Feldman and DiScipio, 1972; MacDonald and Butler, 1974). Each paper presents case study data. Feldman and DiScipio describe a case in which the combined use of behavior therapy (in the form of systematic desensitization) and physical therapy produced positive results in a woman with Parkinson's disease. Prior to intervention, she refused to become involved in activities involving ambulation because of genuine physical difficulties combined with a fear of falling. Once the fear of falling had been alleviated by means of desensitization, ambulation was possible with the use of a walking-aid. The other two studies report the combined use of positive reinforcement for walking, and the extinction, i.e., ignoring, of behaviors associated with helplessness; intervention was successful in both studies.

Pain often accompanies physical problems, and this can further interfere in other areas of functioning not directly affected by the physical problem in question. The concept of "operant pain" has proved to be quite useful in conceptualizing and formulating treatment for chronic pain. Behaviors associated with pain (posturing, ingesting pain medication, sleeping) are seen as escape behaviors providing relief from noxious stimulation. Avoidance behaviors may develop, and these may be extremely resistant to extinction. Levendusky and Pankratz (1975) present a case study of a 65-year-old retired Army officer with a history of numerous abdominal operations and who reported intense pain, loss of weight, and social withdrawal. The authors, using an approach developed by Fordyce, Fowler, Lehman, and DeLateur (1968, 1973), were able to reduce substantially behavior involving extensive use of pain medication and awkward and embarrassing posturing. Pain medication was eventually eliminated by surreptitiously introducing a placebo, and later debriefing the patient.

The essential components of behavioral pain therapy, as described by Fordyce et al., are rearranging the contingencies for "pain" behavior; training in relaxation methods; the social reinforcement of "well" talk; the use of imagery in which pain is regarded as controllable by the patient; and time-contingent, rather than pain-contingent, medication schedules. Using a different approach, Melzack and Perry (1975) report success in all age-groups

with the use of EEG alpha-wave biofeedback training in combination
with hypnosis to decrease reported pain. Alpha conditioning alone
was ineffective. The authors speculated that distraction from
pain, and the acquisition of a sense of personal control, contributed to the positive outcome.

A variety of problems presenting as chronic medical conditions
have been modified by behavioral techniques, but reported gains
have generally been quite modest. Breeden and Kondo (1975) report
limited success with biofeedback in the treatment of left-sided
torticollis. When the therapeutic gains were brought to the female
patient's attention, they proved transitory. Benson, Alexander,
and Feldman (1975) worked with eleven patients, ranging in age
from 48 to 70, suffering from premature ventricular contractions
(PVC's); all were ambulatory, with confirmed ischemic heart
disease. Relaxation training was associated with decreased PVC's
in eight of the patients.

Using an avoidance conditioning paradigm, Malament, Dunn, and
Davis (1975) addressed the problem of wheel-chair patients' pressure sores, which arise from continuous sitting or lying. Pressure
may be relieved by having the patient lift the body with the arms.
When instructions were given to perform these push-ups periodi-
cally, no increase in this behavior was noted. With the addition
of a 30-second alarm contingent on failure to perform a push-up of
at least 45-seconds' duration within each 10-minute period, a
gradual increase in the desired behavior was observed. This was
maintained even after the alarm was disconnected.

Although some of the directions reported in this area are seen
as promising, many of the reports--particularly the single case-
studies--are open to a variety of interpretations. There is an
unfortunate lack of experimental studies in which appropriate control conditions or reversal designs are employed.

HYGIENE, SELF-CARE, AND SELF-MANAGEMENT

As with physical problems, the problems of hygiene, self-care
and self-management are more likely to be brought to the attention
of the behavior therapist by hospital or nursing-home staff than
by the patients themselves. The need for assistance in feeding
and self-care, self-injurious behavior, problems in consumption,
incontinence, and aggressive behaviors are all difficulties which
interfere with the smooth operation of an institution, but are
unlikely to be brought to clinical attention by self-referral.

Two available studies indicate the feasibility of using operant procedures to reinstate self-feeding. Baltes and Zerbe (1976)
report the results of a single-subject study in which an ABAB

reversal design was planned. The subject was a 67-year-old woman
who had shown no self-maintenance behavior for a period of five
months. Initial base-line data revealed near-zero frequency of
self-feeding. This increased dramatically during the treatment
phase, in which a package of treatment techniques was used: Verbal
prompting, stimulus control procedures, immediate reinforcement
with other tangible reinforcers, and a time-out procedure which
consisted of the experimenter turning her back when the patient
refused to eat. Baseline conditions were re-established, with a
concomitant decrease in self-feeding. The patient died before the
treatment condition could be reinstated. Geiger and Johnson (1974),
working with six geriatric patients with low rates of appropriate
eating, were able to increase the number of meals eaten "correctly"
from 12% to 84%. The technique consisted of presentation of a pre-
viously-chosen gift for appropriately eating the evening meal. An
ABAB reversal design was also used in this study. For three of
the patients, "correct" eating was strongly associated with treatment, and incorrect eating with the baseline condition. Patient 4
increased correct eating but failed to show a reversal. Patients
5 and 6 left the facility before completion of the experiment.

Re-establishment of self-care behavior has been reported in
studies in which operant techniques were used; the only populations
for whom data are reported are, however, institutionalized groups,
and this supports the claim made earlier that institutional staff
are most likely to raise this kind of problem for clinical attention. Filer and O'Connell (1964) report an increase in functional
levels of personal appearance and hygiene when chronic Veterans'
Administration Hospital patients were given differential reinforcement. The performance of these subjects was significantly better
than that of a control group with similar expectancies but minimal
reinforcement. Gottfried and Verdicchio (1974) were able to reinstitute bedmaking, teeth-brushing, face-washing and shaving in
50 chronic psychotic patients with the use of only tobacco or
candy as reinforcers. Numerous other programs based on the token
economy (cf. Ayllon and Azrin, 1968) have included older, more
chronic patients and have reported success in re-establishing self-
care behaviors.

Behavioral approaches have been used widely, but with no great
sophistication, for the treatment of incontinence in geriatric
patients. Grosicki (1968) and Pollock and Liberman (1974) report
negative results. In the latter study, correct toileting was
never specifically reinforced; the behavior that was reinforced
was "being dry." Given the lack of specificity in this "response,"
it is hardly surprising that conditioning was not evidenced. In
the Grosicki study, neither social reinforcement nor tangible
reinforcement produced improvement. However, other studies using
operant procedures report positive outcomes (Atthowe, 1972; Carpenter and Simon, 1960; Wagner and Paul, 1970). Collins and Plaska

(1975) used Mowrer's "bell and blanket" method for treating enuresis, and found a significant reduction in the weekly frequency of
bedwetting in the course of eight weeks of treatment. In general,
the more sophisticated and intensive the treatment program, and
the less serious the problem, the better the prognosis. The type
of intensive program developed by Azrin, Sneed, and Foxx (1973)
could well be applied to an incontinent geriatric population.

Kastenbaum and Mishara (1971) and Mishara and Kastenbaum
(1973) have reported a high incidence of "self-injurious" behavior,
but this term is defined broadly to include failure to eat, failure to dress, ingesting inedible objects, and medically-contraindi-
cated overactivity. Mishara, Robertson, and Kastenbaum (1973) present a five-step approach to modification of such behaviors:
Observation, reflection, intervention, assessment, and follow-up.
Observation entails obtaining base-rates of the behaviors in question and collecting data to determine possible relationships with
antecedent or consequent conditions. Reflection refers to an
evaluation process in which questions of values, as well as available resources, are considered before a decision is made to inter-
vene. Intervention consists largely of operant extinction and
reinforcement of competing or incompatible behaviors. Assessment
and follow-up are presented as important considerations, but no
guidelines are given.

No research appears to be available on the use of behavioral
techniques for the treatment of substance abuse in the elderly,
although cigarette-smoking has been targeted. No studies have exclusively used an older population, but several have included
older subjects in their samples (Keutzer, 1968; Ochsner and Damrau,
1970; Wagner and Bragg, 1970; Whitman, 1969, 1972). A variety of
techniques including aversive conditioning, self-control and
coverant-control procedures, negative practice, and systematic desensitization have produced positive results, with no single technique showing clear superiority to the others. In studies in
which age was examined as a possible predictor of outcome (Keutzer,
1968; Wagner and Bragg, 1970), no differential results were
obtained.

COGNITIVE FUNCTIONING AND EMOTIONALITY

Assessment and Theoretical Issues

In this section, the term "emotionality" refers to the affective domain of human experience and behavior: The observations of
a patient's functioning that would be listed under "affect" in a
clinician's mental status examination, or the patient's report of
anxiety, depression, anger and so forth. "Cognitive functioning,"

however, embraces two concepts: First, the cognitive abilities of
the individual in terms of intellectual functioning (attention
span, short- and long-term memory, computation skills, etc.), and,
second, the cognitive "content" of the individual in terms of self-
defeating thoughts, intrusive images, etc., in the sense used in
"cognitive therapy" (Beck, 1976); the "cognitive modality" (Lazarus,
1976); and "cognitive behavior modification" (Meichenbaum, 1974).
Unfortunately, these distinctions are made somewhat unnecessarily
in view of the regrettable paucity of material relevant to this
review.

Some recent general articles conclude optimistically. Jarvik
(1976) provides a thoughtful assessment of depression in the aged,
covering some manifestations of depression in the elderly which are
biochemically atypical, and commenting on the availability of promising therapeutic techniques ranging from antidepressant medication
to focus on the therapeutic milieu. Similarly, Fracchia, Sheppard,
and Merlis (1974) point to some unusual features of the psychological functioning of elderly psychiatric patients, and conclude in
favor of a behavior modification approach to treatment. Inskip
(1970) provides a speculative, but enthusiastic, paper which attempts to counter therapeutic defeatism when confronted by patients
diagnosed as suffering from "chronic brain syndrome;" and Tobias
and MacDonald (1974), after a comprehensive review, conclude that
withdrawal or reduction of maintenance drugs used with long-term
hospitalized patients may be beneficial because of the inhibitory
effect upon learning of higher doses of chlorpromazine, for example.

The controversy over the etiology of intellectual decline in
the elderly has been recently reviewed by Elias, Elias, and Elias
(1977). Some of the apparent decline in intellectual abilities
with increased age may be attributable to situational and attitudinal factors, rather than to a reduction in capacity per se.
For example, Birkhill and Schaie (1975) found that performance on
four of the five factors of the Primary Mental Abilities test could
be enhanced significantly by altering key situational factors in
the standard form of administration of the tests. It is unknown,
however, whether such contingencies would equally benefit the
young, as their sample included only older adults. Hoyer, Labouvie,
and Baltes (1973) found that although response rates of older
adults on speeded tasks were improved through practice, no differential benefit due to reinforcement contingencies was obtained.
Also, no generalization to performance on tests of intellectual
ability was achieved. Erber (1976) also demonstrated the beneficial effect of practice on the Digit Symbol subtest of the WAIS,
one of the most difficult for older adults. However, the practice
effect was comparable for both young and old and when reinforcement
contingencies were added to the experimental design (Grant, Storandt,
and Botwinick, in press), no differential benefits of incentive were
noted for either age group.

Meichenbaum's (1974) self-instructional strategy training represents a significant new development, at least theoretically at the present. The procedure, well documented with younger patients and populations (Meichenbaum, 1974), consists of (1) analyzing the series of accessible cognitive processes involved in a problem-solving task; (2) translating these into specific self-statements; and (3) modeling these and having the patient rehearse them.

It seems clear that a behavioral approach to problems of later life in the affective and cognitive domains holds significant promise; it is equally clear that this promise is, at present, largely unfulfilled.

Individual and Group Treatment

In comprehensive reviews of the literature of group psychotherapy, reports of applications to the elderly are hard to find (Lubin, Lubin, and Sargent, 1972; MacLennan and Levy, 1969). One good example is the report by Weinman et al (1972) noted earlier, if assertive responses are considered to fall within the "emotional" domain. Reported token economies, in the main, focus upon social, self-care, and occupational behaviors; where cognitive and emotional functioning are assessed, typically no beneficial changes are found (Fraser et al, 1976; Mumford et al, 1975). An intriguing study by Powell (1974) showed that a program of mild exercise actually increased scores on intellectual tests (Standard Progressive Matrices and Wechsler Memory Scale) more than "social therapy" or no treatment. Perhaps this finding supports the utility of Lazarus' (1976) concept of "multimodal" assessment and therapeutic endeavors.

In a case-study which seems to demonstrate the value of a multi-faceted behavioral treatment approach, Flannery (1974) successfully treated a 77-year-old man with physical and social complaints apparently related to grief by means of behavioral contracting, in which therapeutic sessions were contingent upon the patient's following a five-step contract; by the social reinforcement of statements reflecting a positive self-evaluation; and by prompting, and shaping, the discussion of grief-related topics. This approach is refreshingly contrasted with the rather speculative reports and clinical descriptions of grief which are prevalent in this literature (e.g. Burnside, 1969).

In a recent report of a cognitive-behavioral analysis of grief, Gauthier and Marshall (1977) speculated that two factors could be involved in unreasonably protracted or excessive grief: First, the social reinforcement (by family, friends, etc.) of grief behaviors in the form of solicitousness and sympathy; and second, the "con-

spiracy of silence," in which the bereaved is carefully protected
from all reminders of the loved departed. This latter could result
in the "cognitive incubation" of grief responses. In a series of
four patients with chronic grief (the ages of the patients were
not specified), the authors reported marked success in only half-
a-dozen treatment sessions. The "significant others" of each
patient were taught to restructure their responses to the patient's
grieving behavior (giving attention when the patient behaved otherwise than showing grief, and being relatively non-sympathetic to
grief-related depression); and the patient received "flooding"
sessions in which he/she was asked to confront, as vividly as
possible, imagery relating to the deceased. Not only the relatives, but the patients themselves were reported to be extremely
pleased by the outcome.

MODIFICATION OF STAFF BEHAVIOR AND ATTITUDES

In evaluating treatment approaches to the elderly in institutional settings (i.e. nursing homes or psychiatric hospitals), two
distinctions made earlier have implications for the desirability
of modifying staff behavior: The distinction between behaviors
seen as problematic by the staff, and those complained of by the
client/patient; and the distinction between client problems attributable to the effects of aging or of specific psychiatric/neuro-
logical syndromes, and those attributable to the prevailing contingencies of the social milieu of the institution. The question
may be raised with either distinction: "Whose behavior needs to
be modified--the clients' or the staff's?" [In a positive vein,
the enormous therapeutic potential of the psychiatric nurse has
been the subject of a recent conference in Britain (Marks, Peck,
Kolvin, Hall, Rosenthall, Kiernan, and Connolly, 1973) the conclusions of which were extremely optimistic.] It is thus very
appropriate that behavior therapy techniques have been successfully
applied to the behavior and attitudes of those who care for the
elderly in institutional settings.

In a study of general relevance to this theme, Cautela and
Wisocki (1969) asked students to fantasize having their lives saved
by an elderly person, and to practice this imagery procedure at
home during a ten-day period. The students subsequently evidenced
a significantly more favorable attitude toward the elderly on a
questionnaire measure than at a previous administration of the
measure, whereas a similar group who had not had the imagery exercises showed no change in attitude. Further examination of the
data showed that the students who claimed to have practiced the
imagery exercise more often showed significantly more positive
change in attitude than those who practiced less. Cautela and
Wisocki hypothesized that imaginally pairing an elderly person with
a "reinforcing" situation would produce favorable attitude change

by counterconditioning. Limitations of this study are that there
was no control for the possible demand characteristics of the situation at the second administration of the questionnaire, and that
results are limited to questionnaire responses on two occasions
separated by about 2⅔ weeks.

McReynolds and Coleman (1972) called attention to the neglect
of staff attitudes as an important variable in the success of a
token economy. In the context of a report on a successful token
economy program with severely regressed institutionalized psychiatric patients, the authors found that improvement in the behavior
of the patients was accompanied by an increase in staff enthusiasm
and optimism. This "cognitive behavior modification" seemed attributable to the training seminars available in the program, and
to staff perception of the positive results of the program.

In an ambitious program in which staff behavior and attitudes
towards geriatric patients were actively and explicitly targeted,
Hickey (1974) attempted an intensive and multifaceted training program for the staff of a hospital geriatrics ward. Essentials of
the program were that all direct-service personnel were taught in
the ward context; training took specific account of individual
staff members' position in the hospital hierarchy; and, in a small-
group format, learning or problem-solving efforts replaced the
conventional teaching model. Over a five-month period, 130 staff
participated in 18 two-hour sessions which included role-playing,
film presentations, and practical exercises. Changes in staff behavior and attitudes were assessed by means of several measures.
Conclusions from self-report questionnaires were that the intensive, on-site approach was effective; that paraprofessional staff
gained most benefit from the program, at least in terms of their
reported satisfaction; and that direct-care staff participated more
enthusiastically in the "active" procedures (e.g. role-playing,
psychodrama) than in the "passive" approaches (didactic presentations, etc.). Other general findings were that staff turnover was
reduced; the number of patients returned to the community doubled;
and there were beneficial administrative changes on the unit.

In similar vein, Heller and Walsh (1976) attempted to influence positively the attitudes of nursing students towards the
elderly in a training program, the results of which support some of
the findings of Hickey (1974). The broad-ranging program centered
around experiential learning, which included role-playing and psychodrama sessions. A battery of relevant questionnaire indices
showed significantly more favorable attitudes towards gerontology
as a specialization after the course as compared to preliminary
assessment. No change in attitude was found in a "control" group
(whose activities were unspecified), in spite of the fact that the
control students acquired relevant information from the experimental subjects. Results were interpreted in terms of cognitive

dissonance theory. Although the results may not be applicable to other settings, and the crucial components of the program for fostering attitude change are not discernible, the study makes a good case for supervised exposure to relevant work settings as a means of increasing interest in gerontology.

It appears from the above studies that (1) the successful application of behavioral methods to patients' problems may serve to improve, in turn, staff enthusiasm (which, presumably, may "feed back" beneficially to the patients); and (2) behavior therapy approaches may effectively be directed at staff behavior in institutions, again producing a desired effect on the ward. The effectiveness of behavior therapy in institutions may thus be enhanced by, in a sense, treating both patients and staff.

CONCLUSIONS

(1) Behavior therapy techniques that have been found useful with younger populations seem effective with the elderly, at least in those areas in which applications have been reported.

(2) Behavior therapy with the elderly seems to have been conducted on a low plane of sophistication, in which newer theoretical developments (e.g. biofeedback, cognitive-behavior modification, multimodal behavior therapy) have been practically ignored.

(3) The problems of later life per se have generated no specific treatment techniques.

(4) Many of the problems addressed by behavior therapists in the published reports were not "complained of" by the patients, but were presented as undesirable by the institutional staff.

(5) There are few controlled studies of behavioral procedures with elderly groups.

(6) There is some evidence that it is important to modify behavioral treatment techniques for implementation with the elderly (at least with elderly chronic mental patients).

(7) Many of the research results confound "the elderly" with "the chronically institutionalized patient."

(8) Treatment research for some problem areas has not been reported (e.g. sexual dysfunction; marital disharmony).

(9) Azrin's rapid, conglomerate methods of behavioral treatment have not been applied to the elderly (Azrin and Armstrong, 1973).

(10) Most applications seem to have been inspired by the model of loss, debilitation, decline, and deficit in later life, rather than by an optimistic viewpoint which would generate attempts to foster personal growth and development.

REFERENCES

Atthowe, J. M., Jr. Controlling nocturnal enuresis in severely disabled and chronic patients. Behavior Therapy, 1972, 3, 232–239.

Ayllon, R. and Azrin, N. H. The token economy. New York: Appleton-Century-Crofts, 1968.

Azrin, H. H. and Armstrong, P. M. The "mini-meal"--A method for teaching eating skills to the profoundly retarded. Mental Retardation, 1973, 2, 9–13.

Azrin, N. H., Sneed, T. J., and Foxx, R. M. Dry bed: A rapid method of eliminating bedwetting (enuresis) of the retarded. Behavior Research and Therapy, 1973, 11, 427–434.

Baker, R. The use of operant conditioning to reinstate speech in mute schizophrenics. Behavior Research and Therapy, 1971, 9, 329–336.

Baltes, M. M. and Lascomb, S. L. Creating a healthy institutional environment for the elderly via behavior management: The nurse as a change agent. International Journal of Nursing Studies, 1975, 12, 5–12.

Baltes, M. M. and Zerbe, M. H. Reestablishing self-feeding in a nursing home resident. Nursing Research, 1976, 25, 24–26.

Bandura, A. Principles of behavior modification. New York: Holt, Rinehart & Winston, 1969.

Barnes, E. K., Sack, A., and Shore, H. Guidelines to treatment approaches: Modalities and methods for use with the aged. The Gerontologist, 1973, 13, 513–527.

Beck, A. T. Cognitive therapy and the emotional disorders. New York: International Universities Press, 1976.

Benson, H., Alexander, S., and Feldman, C. L. Decreased premature ventricular contractions through use of the relaxation response in patients with stable ischaemic heart-disease. Lancet, 1975, 2(7931), 380–382.

Birkhill, W. R. and Schaie, K. W. The effect of differential reinforcement of cautiousness in intellectual performance among the elderly. Journal of Gerontology, 1975, 30, 578–585.

Bollinger, R. L. Geriatric speech pathology. The Gerontologist, 1974, 14, 217–220.

Breeden, S. A. and Kondo, D. Using biofeedback to reduce tension. American Journal of Nursing, 1975, 75, 2010–2012.

Brody, E. M., Cole, C., and Moss, M. Individualizing therapy for the mentally impaired aged. Social Casework, October 1973, 453–461.

Burnside, I. M. Grief work in the aged patient. Nursing Forum,
    1969, 8, 417–427.
Burt, B., Law, M., Machan, L., Macklin, C., Nesbitt, D., Read, H.,
    and Wiebe, C. Mildred Jones walks again. Canadian Nurse,
    1974, 70, 37.
Butler, R. N. and Lewis, M. I. Aging and mental health (2nd ed.).
    St. Louis: C. V. Mosby, 1977.
Carpenter, H. A. and Simon, R. The effect of several methods of
    training on long-term, incontinent, behaviorally regressed
    hospitalized psychiatric patients. Nursing Research, 1960,
    9, 17–22.
Cautela, J. R. Behavior therapy and geriatrics. Journal of Gene-
    tic Psychology, 1966, 108, 9–17.
Cautela, J. R. A classical conditioning approach to the develop-
    ment and modification of behavior in the aged. The Gerontolo-
    gist, 1969, 9, 109–113.
Cautela, J. R. and Mansfield, L. A behavioral approach to geri-
    atrics. In W. D. Gentry (Ed.) Geropsychology: A model of
    training and clinical service. Cambridge: Ballinger, 1977.
Cautela, J. R. and Wisocki, P. A. The use of imagery in the
    modification of attitudes toward the elderly: A preliminary
    report. Journal of Psychology, 1969, 73, 193–199.
Cohen, D. Research problems and concepts in the study of aging:
    Assessment and behavioral modification. The Gerontologist,
    1967, 7, 13–19.
Collins, R. W. and Plaska, T. Mowrer's conditioning treatment for
    enuresis applied to geriatric residents of a nursing home.
    Behavior Therapy, 1975, 6, 632–638.
Corby, N. Assertion training with aged populations. The Coun-
    seling Psychologist, 1975, 5, 69–74.
Davison, G. C. and Stuart, R. B. A statement on behavior modifi-
    cation from the Association for the Advancement of Behavior
    Therapy. AABT Newsletter, May 1974, 2–3.
Elias, M. F., Elias, P. K., and Elias, J. W. Basic processes in
    adult developmental psychology. St. Louis: C. V. Mosby Co.,
    1977.
Erber, J. Age differences in learning and memory of a Digit-Symbol
    substitution task. Experimental Aging Research, 1976, 2,
    45–53.
Eysenck, H. J. and Rachman, S. The causes and cures of neurosis.
    London: Routledge and Kegan Paul, 1965.
Feldman, M. G. and DiScipio, W. J. Integrating physical therapy
    with behavior therapy. Physical Therapy, 1972, 52, 1283–1285.
Filer, R. N. and O'Connell, D. D. Modification of aging persons
    in an institutional setting. Journal of Gerontology, 1964,
    19, 15–22.
Flannery, R. B. Behavior modification of geriatric grief: A trans-
    actional perspective. International Journal of Aging and
    Human Development, 1974, 5, 197–203.

Fordyce, W. E., Fowler, R. S., Lehmann, J. F., and DeLateur, B.
    Some implications of learning in problems of chronic pain.
    Journal of Chronic Disease, 1968, 21, 179–190.
Fordyce, W. E., Fowler, R. S., Lehmann, J. F., and DeLateur, B.
    Operant conditioning in the treatment of chronic pain.
    Archives of Physical Medicine and Rehabilitation, 1973, 54,
    399–408.
Fracchia, J., Sheppard, C., and Merlis, S. Psychological charac-
    teristics of long-term mental patients: Some implications for
    treatment. Comprehensive Psychiatry, 1974, 15, 495–501.
Franks, C. M. (Ed.). Behavior therapy: Appraisal and status. New
    York: McGraw-Hill, 1969.
Franks, C. M. and Wilson, G. T. (Eds.). Annual review of behavior
    therapy: Theory and practice (Vol. I). New York: Brunner/
    Mazel, 1973.
Franks, C. M. and Wilson, G. T. (Eds.). Annual review of behavior
    therapy: Theory and practice (Vol. II). New York: Brunner/
    Mazel, 1974.
Franks, C. M. and Wilson, G. T. (Eds.). Annual review of behavior
    therapy: Theory and practice (Vol. III). New York: Brunner/
    Mazel, 1975.
Franks, C. M. and Wilson, G. T. (Eds.). Annual review of behavior
    therapy: Theory and practice (Vol. IV). New York: Brunner/
    Mazel, 1976.
Fraser, D., McLeod, W. L., Begg, J. C., Hawthorne, J. H., and Davis,
    P. Against the odds: The results of a token economy pro-
    gramme with long-term psychiatric patients. International
    Journal of Nursing Studies, 1976, 13, 55–63.
Gauthier, J. and Marshall, W. L. Grief: A cognitive-behavioral
    analysis. Cognitive Therapy and Research, 1977, 1, 39–44.
Geiger, O. G. and Johnson, L. A. Positive education for elderly
    persons: Correct eating through reinforcement. The Geron-
    tologist, 1974, 14, 432–436.
Gottesman, L. E. Clinical psychology and aging: A role model.
    In W. D. Gentry (Ed.). Geropsychology: A model of training
    and clinical service. Cambridge: Ballinger, 1977.
Gottfried, A. W. and Verdicchio, F. G. Modifications of hygienic
    behaviors using reinforcement therapy. American Journal of
    Psychotherapy, 1974, 28, 122–128.
Grant, E. A., Storandt, M. and Botwinick, J. Incentive and prac-
    tice in the psychomotor performance of the elderly. Journal
    of Gerontology, in press.
Grosicki, J. P. Effect of operant conditioning on modification of
    incontinence in neuropsychiatric geriatric patients. Nursing
    Research, 1968, 17, 304–311.
Group for the Advancement of Psychiatry. The aged and community
    mental health: A guide to program development, Vol. 8, series
    No. 81, Nov. 1971.
Hampe, G. O. and Blevins, A. L. Primary group interaction of resi-
    dents in a retirement hotel. International Journal of Aging

and Human Development, 1975, 6, 309-320.

Heller, B. R. and Walsh, F. J. Changing nursing students' attitudes toward the aged: An experimental study. Journal of Nursing Education, 1976, 15, 9-17.

Hickey, T. In-service training in gerontology. The Gerontologist, 1974, 14, 57-64.

Hoyer, W. J. Application of operant techniques to the modification of elderly behavior. The Gerontologist, 1973, 13, 18-22.

Hoyer, W. J., Kafer, R. A., Simpson, S. C., and Hoyer, F. W. Reinstatement of verbal behavior in elderly mental patients using operant procedures. The Gerontologist, 1974, 14, 149-152.

Hoyer, W. J., Labouvie, G. V., and Baltes, P. B. Modification of response speed deficits and intellectual performance in the elderly. Human Development, 1973, 16, 233-242.

Hoyer, W. J., Mishara, B. L., and Riedel, R. Problem behaviors as operants. The Gerontologist, 1975, 15, 452-465.

Hutchison, S. L. An investigation of learning under two types of social reinforcers in young and elderly adults. International Journal of Aging and Human Development, 1974, 5, 181-186.

Inskip, W. Treatment programs for patients with chronic brain syndrome can be successful. Journal of the American Geriatrics Society, 1970, 18, 631-636.

Jarvik, L. F. Aging and depression: Some unanswered questions. Journal of Gerontology, 1976, 31, 324-326.

Kahn, R. L. Perspectives in the evaluation of psychological mental health programs for the aged. In W. D. Gentry (Ed.) Geropsychology: A model of training and clinical service. Cambridge: Ballinger, 1977.

Kastenbaum, R. and Mishara, B. L. Premature death and self-injurious behavior in old age. Geriatrics, 1971, 26, 70-91.

Keutzer, C. S. Behavior modification of smoking: The experimental investigation of diverse techniques. Behavior Research and Therapy, 1968, 6, 137-157.

Kleban, M. H., Lawton, M. P., Brody, E. M., and Moss, M. Behavioral observations of mentally impaired aged: Those who decline and those who do not. Journal of Gerontology, 1976, 31, 333-339.

Lawton, M. P. and Gottesman, L. E. Psychological services to the elderly. American Psychologist, 1974, 29, 689-693.

Lazarus, A. A. Behavior therapy and beyond. New York: McGraw-Hill, 1971.

Lazarus, A. A. Multimodal behavior therapy. New York: Springer, 1976.

Levendusky, P. and Pankratz, L. Self-control techniques as an alternative to pain medication. Journal of Abnormal Psychology, 1975, 84, 165-168.

Leviton, D. The significance of sexuality as a deterrent to suicide among the aged. Omega, 1973, 4, 163-174.

Lindsley, O. R. Geriatric behavioral prosthetics. In R. Kasten-
    baum (Ed.). New thoughts on old age. New York: Springer,
    1964.

Lubin, B., Lubin, A. W., and Sargent, C. W. The group psycho-
    therapy literature, 1971. International Journal of Group
    Psychotherapy, 1972, 22, 492-529.

MacDonald, M. L. and Butler, A. K. Reversal of helplessness:
    Producing walking behavior in nursing home wheelchair resi-
    dents using behavior modification procedures. Journal of
    Gerontology, 1974, 29, 97-101.

MacLennan, B. W. and Levy, N. The group psychotherapy literature,
    1968. International Journal of Group Psychotherapy, 1969,
    19, 382-400.

McReynolds, W. T. and Coleman, J. Token economy: Patient and
    staff changes. Behavior Research and Therapy, 1972, 10,
    29-34.

Malament, I. R., Dunn, M. E., and Davis, R. Pressure sores: An
    operant conditioning approach to prevention. Archieves of
    Physical Medicine and Rehabilitation, 1975, 56, 161-164.

Marks, I., Peck, D. F., Kolvin, I., Hall, J. N., Rosenthall, G.,
    Kiernan, C. and Connolly, J. The psychiatric nurse as a
    therapist. Nursing Times, Nov. 1973, (supplement).

Meichenbaum, D. H. Cognitive behavior modification. Morristown:
    General Learning Press, 1974.

Meichenbaum, D. H. Self-instructional training: A cognitive
    prosthesis for the aged. Human Development, 1974, 17,
    273-280.

Melzak, R. and Perry, C. Self-regulation of pain: The use of
    alpha-feedback and hypnotic training for the control of
    chronic pain. Experimental Neurology, 1975, 46, 452-469.

Mishara, B. L. and Kastenbaum, R. Self-injurious behavior and
    environmental change in the institutionalized elderly. Inter-
    national Journal of Aging and Human Development, 1973, 4,
    133-145.

Mishara, B. L., Robertson, B., and Kastenbaum, R. Self-injurious
    behavior in the elderly. The Gerontologist, 1973, 6, 311-314.

Mumford, S. J., Patch, I. C. L., Andrew, N., and Wyner, L. A
    token economy ward programme with chronic schizophrenic
    patients. British Journal of Psychiatry, 1975, 126, 60-72.

Ochsner, A. and Damrau, F. Control of cigarette habit by psycho-
    logical aversive conditioning: Clinical evaluation in 53
    smokers. Journal of the American Geriatrics Society, 1970,
    18, 365-369.

O'Leary, K. D. and Wilson, G. T. Behavior therapy: Application
    and outcome. Englewood Cliffs: Prentice Hall, 1975.

Olson, R. P. and Greenberg, D. J. Effects of contingency con-
    tracting and decision-making groups with chronic mental
    patients. Journal of Consulting and Clinical Psychology,
    1972, 38, 376-383.

Paul, G. L.   Chronic mental patient: Current status--future direc-
    tions. Psychological Bulletin, 1969, 71, 81-94.
Peck, D. F.   Behaviour is behaviour is behaviour.   Paper presented
    at the Second European Conference on Behavior Modification,
    Wexford, Ireland, Sept. 1972.
Peck, D. F.   Operant conditioning and physical rehabilitation.
    European Journal of Behavioural Analysis and Modification,
    1976, 3, 158-164.
Pollock, D. D. and Liberman, R. P.   Behavior therapy of inconti-
    nence in demented inpatients.   The Gerontologist, 1974, 13,
    488-491.
Powell, R. R.   Psychological effects of exercise therapy upon
    institutionalized geriatric mental patients.   Journal of
    Gerontology, 1974, 29, 157-161.
Preston, C. F.   Behavior modification: A therapeutic approach to
    aging and dying.   Postgraduate Medicine, 1973, 54, 64-68.
Richard, J., Picot, A., deBus, P., Andreoli, A., and Dalakaki, X.
    De l'application de la relaxation en geriatrie hospitaliere.
    Archives Suisses de Neurologie, Neurochirurgie et Psychiatrie,
    1975, 117, 157-169.
Riedel, R. G.   Experimental analysis as applied to adulthood and
    old age: A review.   Paper presented at the meeting of the
    American Psychological Association, New Orleans, Aug. 1974.
Rimm, D. C. and Masters, J. C.   Behavior therapy: Techniques and
    empirical findings.   New York: Academic Press, 1974.
Roth, M.   The psychiatric disorders of later life.   Psychiatric
    Annals, 1976, 6, 417-445.
Sabatasso, A. P. and Jacobson, L. I.   Use of behavioral therapy
    in the reinstatement of verbal behavior in a mute psychotic
    with chronic brain syndrome: A case study.   Journal of
    Abnormal Psychology, 1970, 76, 322-324.
Seligman, M. E. P.   Helplessness: On depression, development, and
    death.   San Francisco: W. H. Freeman and Co., 1975.
Tobias, L. L. and MacDonald, M. L.   Withdrawal of maintenance drugs
    with long-term hospitalized mental patients: A critical
    review.   Psychological Bulletin, 1974, 83, 107-125.
Wagner, M. K. and Bragg, R. A.   Comparing behavior modification
    approaches to habit decrement--smoking.   Journal of Consult-
    ing and Clinical Psychology, 1970, 34, 258-263.
Wagner, B. R. and Paul, G. L.   Reduction in incontinence in chronic
    mental patients: A pilot project.   Journal of Behavior Ther-
    apy and Experimental Psychiatry, 1970, 1, 29-38.
Weinman, B., Gelbart, P., Wallace, M., and Post, M.   Inducing
    assertive behavior in chronic schizophrenics: A comparison
    of socioenvironmental, desensitization and relaxation ther-
    apies.   Journal of Consulting and Clinical Psychology, 1972,
    39, 246-252.
Whitman, T.   Modification of chronic smoking behavior: A compari-
    son of three approaches.   Behavior Research and Therapy, 1969,
    7, 257-263.

Whitman, T. L.   Aversive control of smoking behavior in a group
    context.   Behavior Research and Therapy, 1972, 10, 97–104.
Wolpe, J. and Lazarus, A. A.   Behavior therapy techniques.   New
    York: Pergamon Press, 1966.

OTHER APPROACHES TO THERAPY

Martha Storandt

Washington University

This chapter represents a _pot pourri_ of techniques, ideas,
frames of reference and "therapies," each of which, in and of
itself, has not been sufficiently dealt with by the gerontological
research community to warrant treatment in a separate chapter.
This is not to imply that these treatment techniques are not
widely applied if we consider all therapies currently employed
with older adults. Large numbers of institutions caring for the
mentally ill aged devote many staff-hours to remotivation therapy.
Many group therapists may use a psychodynamic frame of reference
in their treatment of older clients. However, relatively little
research presently exists with respect to the application of these
therapeutic techniques to older adults. Many of the articles and
reports which can be found in the literature are think pieces,
program descriptions, case studies, or recommendations based on a
subjective evaluation on the part of the practitioner. Thus, the
research clinical psychologist, with the specialized skills which
training in clinical psychology produces, may find this chapter
a veritable goldmine of intriguing hypotheses; on the other hand,
relatively little in the way of "facts," or even substantiated
theory, will be apparent.

GROUP THERAPY

Many of the other approaches to therapy with the aged described in this chapter involve groups of older adults, e.g.,
remotivation, reality orientation, art therapy. Group therapy
has been employed with both outpatient older adults and the institutionalized aged, as well as with groups of mixed ages. The

group may focus upon brief, crisis oriented psychotherapy (Oberleder, 1970) or may be part of a larger, ward-wide activity program (Reichenfeld, Csapo, Carriere, and Gardner, 1973).

Little good quality research exists which examines the effectiveness of traditional group therapy with older adults in terms
of the patient characteristics leading to greatest benefits.
Further, group therapy with old, as compared to young, patients
has not been investigated. It may well be that the character of
effective traditional group techniques, of any persuasion, must be
modified when the group members are older adults. For example,
members of institutionalized groups of older adults frequently
come and go, sometimes making it difficult to develop the kind of
group dynamics found in traditional groups. The course of therapy
may be longer when the group contains older outpatients (Liederman
and Liederman, 1967). Practical problems of everyday living, such
as housing, transportation, and nutrition, may have to be dealt
with first before interpersonal issues and dynamics can be addressed.

The positive relationship between the presence of a confidant
and the mental health of the older individual (Lowenthal and Haven,
1968) would suggest that opportunities to share with other group
members one's innermost concerns, worries, cares, and joys should
be beneficial to the older adult who has all too often lost the
social and interpersonal roles common to earlier periods of life.
However, the objective appraisal of this hypothesis awaits the
attention of clinical psychology.

MILIEU THERAPY

Milieu therapy was originally developed in Europe and England
as an alternative to the widespread notion that proper care for
mental patients involved incarceration and/or treatment by one or
a few identified therapists. Under the tenets of milieu therapy,
interpersonal relations existing throughout the entire environment are thought to have therapeutic potential. For example, a
brief interaction between a patient and a member of the house-
keeping staff can be used to enhance ego growth on the part of
the patient. Therefore major emphasis is placed upon action, as
opposed to insight, and upon the development of instrumental and
social skills. The application of milieu therapy to the aged has
primarily been in the form of rehabilitation programs designed to
reduce institutionalization responses on the part of chronic
geriatric patients or to assist the older patient in dealing with
crises connected with the changes in social roles which frequently
accompany old age.

A series of studies begun in the early 1960s by the University of Michigan's Institute of Gerontology have described the effects of a therapeutic milieu upon elderly state hospital patients. The principles of the therapeutic ward milieu involved in these studies were those of sex-integrated living conditions in which the patients were allowed and expected to maintain a high degree of decision making and self-management. The program emphasized the importance of work in terms of a sheltered workshop as well as a ward store in which patients could use money earned in the workshop to purchase desired items. The therapeutic milieu also offered frequent opportunities for recreational activities.

One study examined the outcome effects of the therapeutic milieu on newly-admitted geriatric patients (Gottesman, 1965). The treated patients were less docile and passive than controls and were found to be improved in terms of mental status and self-concept. Similar benefits of the therapeutic milieu were observed in chronic geriatric patients who had been hospitalized an average of 18 years (Gottesman, 1967) and in older chronic schizophrenics (Gutmann, Gottesman and Tessler, 1973). "Total push" milieu therapy rapidly improved the social functioning, personal self-care, and instrumentality of psychiatric-medically infirm men (patients with a primary diagnosis of schizophrenia as well as a physical illness such as diabetes) but was effective only for those who lived on closed or semi-closed wards, i.e., those who were seen by the staff as being the most regressed. No change was seen in the behavior of less severely regressed patients, although this lack of improvement may have been due to a ceiling effect in the measurement instrument (Steer and Boger, 1975).

A major study of socio-environmental rehabilitation programs for chronic psychotics of all ages in a state hospital setting (Sanders, Smith, and Weinman, 1967) revealed that such milieu programs are most effective for a subset of older, rather than younger, patients. These include those older adults with manifest psychotic symptomology but with relatively intact cognitive capabilities. The older adult's reduced reactivity (emotional intensity or drive level) was seen as aiding, rather than detracting from, psychiatric adjustment and successful socialization, at least in older men.

Mishara (1977) has attempted to differentiate the type of geriatric patient for whom milieu therapy is most beneficial. In his comparison of milieu therapy with a treatment program involving a token economy, 80 male and female patients diagnosed as demonstrating age-associated chronic brain syndrome were randomly assigned to the two programs. Results of a discriminate analysis of 18 predictor variables indicated that improvement after six months' exposure to milieu therapy was most common in the unresponsive, apathetic male patient while improvement in the token economy was associated with a constellation of measures which seemed to describe the less institutionalized, acting-out older adult. Gatz, Siegler, and Dibner (1977) have noted that the milieu therapy community tends to extrude, or transfer out, those patients who exhibit "negative" behaviors. The negative behaviors they describe may well be comparable to those Mishara found unrelated to improvement in a milieu therapy environment but more effectively treated by the operant procedures of a token economy.

One study employed the principles of milieu therapy in a day center on a university campus (Spence, Cohen, and Kowalski, 1975). Elderly patients from a state hospital were transported to the center by bus daily. Three variables which best predicted external placement following completion of the 14 week special program were identified: staff ratings of level of functioning, time orientation, and the extent to which the patient had community contacts. Indicators of the latter included receiving mail or visitors, having a bank account, and the number of times the patient had been able to leave the hospital in the past. Since most programs of milieu therapy have as their ultimate goal the discharge of the patient back to normal community living, a lack of economic and social resources on the part of the patient may serve as a hindrance to such placements unless a wider range of community support systems are established.

There are indeed a number of factors which may minimize the positive impact of milieu therapy (Bok, 1971). The staff, and patients as well, may perceive older adults as incapable of change. Also, positive change may be expected as a function of some external agent's actions, rather than one's own efforts. "Give me a pill to make me well." "Let the administration make the necessary changes to improve the quality of care delivered in this institution and then I'll do my part." The monolithic bureaucracy of large psychiatric institutions may in fact make environment-wide changes difficult. The therapeutic ideology may be difficult to achieve and therapeutic principles may fail to provide guidelines to the solutions of all of the problems which arise in the therapeutic community (Gatz, et al., 1977). For example, the principle of horizontal, as opposed to hierarchical, staff relations is difficult to effect in reality. Or, patients may see the goal of the therapeutic community as that of providing a humane and caring living environment, while staff may concentrate upon discharge. Smyer, Siegler, and Gatz (1976) have pointed out that despite the emphasis, in principle, placed upon performance criteria staff evaluations of patient personality dimensions actually determined successful outcome in the milieu unit they studied. Further, staff sometimes tend to set unrealistic goals for the unit, especially in terms of speed of progress (Gatz, et al., 1977).

Despite the hindrances noted above the weight of evidence to date would point to the potential benefits of milieu therapy for those older adults who represent a substantial proportion of the institutionalized aged--those for whom other treatment techniques may have been ineffective and who have passively accepted terminal institutionalization.

REALITY ORIENTATION

The Geriatric Reality Orientation program (e.g., Taulbee and Folsom, 1966; Folsom, 1967) is a result of the blending of two treatment philosophies--behavior and milieu. It includes the prescription of particular attitudes on the part of the milieu, classes in reality orientation, and the use of operant conditioning procedures for modifying behavior. It represents a program of treatment originally used with younger patients suffering from brain damage and, according to the originators of the therapy, is "ideally suited for the patient with a moderate to severe degree of organic cerebral deficit resulting from arteriosclerosis . . ." (Taulbee and Wright, 1971, p. 72). In fact, patients initially chosen for the reality orientation program at Tuscaloosa VA Hospital were the most chronic, regressed, senile, medically and psychiatrically infirm in the hospital. Based on nurses' observations of 128 patients assigned to the program over a three year period, 80% of those "untreatable" patients were judged as improved.

It is the classes in reality orientation which appear to form the novel aspect of the therapeutic procedure and which have been described in greatest detail. Usually there are only four patients in a basic class, which meets twice a day, seven days a week. In order to achieve consistency the same two nursing assistants are assigned to instruct the classes on a regular basis. Basic personal and current information is presented over and over in order to teach the patient what his name is, where he is, and the date. When this basic information is learned the instructor moves on to other facts--age, hometown, former occupation.

In addition to the verbal interaction between the patient and the nursing assistant-teacher, the teaching process involves the use of a variety of learning aids or props including a reality orientation board, name plates, puzzles, plastic numbers and letters, and the like. As the patient progresses, he is promoted to an advanced class which meets only once a day and where material is given more rapidly, including new information to be learned.

An effort is made to see that the reality training is carried on 24 hours a day by the remaining ward staff and even by family and visitors. Patients are grouped with others similarly in reality orientation training at meals and in sleeping facilities so that the same individuals are in their environment throughout

the day. Name plates at their tables, above their beds and on
their clothing remind them of who they are. Staff are urged to use
the appropriate names of objects in the environment in their interactions with these patients and to repeatedly identify items for
the patient.

Recent attempts to experimentally evaluate the effectiveness
of geriatric reality orientation have produced mixed results and
point to the many variables which may influence the appropriateness
of its application. Barnes (1974) found no statistically significant changes in six patients assigned to a limited program of one
class per day, six days per week for a period of six weeks. His
measurements included a 23-item patient questionnaire and a 5-item
questionnaire completed by the nursing director of the home--mea-
surement instruments with unknown reliability and validity.
Barnes' negative results could have been related to any one of a
number of factors including: a reduced classroom experience
(once a day as opposed to twice) carried out in one corner of a
presumably busy lounge; the omission of the 24-hours, round-the-
clock follow-up and implementation of attitude and milieu therapy; a limited (six weeks) course of therapy; or inappropriate
measuring devices.

More recently Harris and Ivory (1976) examined the effectiveness of reality orientation with female geriatric patients in a
state mental hospital. Based on a pre-post experimental-control
group design, they reported a beneficial effect of a five-month
program of reality orientation upon a number of verbal orientation behaviors (spontaneity of speaking, knowledge of own name and
names of aides and other patients, compliance with a request, and
time orientation) as well as in terms of aide ratings of orientation, reduction of "crazy talk," isolation or withdrawal. No
significant differences were observed between the treatment and
control groups in terms of ratings of ward behavior or in terms
of aide impressions of general appearance, clarity of speech, general cooperativeness, and continence of bowels.

Citrin and Dixon (1977) also have reported the results of a
pre-post treatment-control group design employed to evaluate the
effectiveness of a seven-week program of reality orientation in a
geriatric institution in the midwest. Improvement was noted in
the 12 experimental subjects on a measure of information, but not
in terms of the Geriatric Rating Scale--a measure of ward behavior.
In these three studies all ratings were conducted by treatment
personnel, providing for a major confound in terms of expectancies
and behavior change, as pointed out by Harris and Ivory (1976).

In the studies described above no attempts were made to assess
the various parameters which may be important to the effectiveness,
or ineffectiveness, of reality orientation. It may well be that

the treatment is appropriate for only certain types of geriatric patients, that the classes must be employed along with the round-the-clock follow-up, that positive benefits cannot be demonstrated in less than several months' time, and that there are a whole host of interaction effects among patient, therapist, and environmental characteristics.

Recently, Brook, Degun and Mather (1975) began the slow process of unraveling the complexities of this therapeutic technique by examining the role of the therapist, in comparison to stimulation with reality orientation materials and the classroom environment alone. On the basis of a rating scale which assessed intellectual and social functioning, 18 demented patients were divided into three groups, reflecting varying levels of impairment. Subjects in each level were then divided into experimental and control groups. Experimental subjects were treated to 16 weeks of classroom reality orientation experience with a therapist present. Controls were allowed to use the reality orientation materials but the therapists did not actively encourage them or participate with the patients in such use.

The patients were then rated every two weeks by the nursing staff, which was unaware of group assignments. Improvement in the first two weeks was noted for all patients. However, only those patients actively involved with a therapist continued to improve over the 16 week course of treatment. Control patients returned to essentially base line levels of social and intellectual performance. Greatest improvement was noted for those patients whose initial levels of functioning were moderate to good, as compared to those demonstrating the lowest initial levels of performance.

Although it is premature to generalize on the basis of a single, unreplicated study, it would appear that the beneficial effects of reality orientation are related to the patient-therapist interaction, as is so often the case with many types of therapies. Obviously a next step would be to determine those characteristics of the reality orientation therapist which are most beneficial to the treatment of the demented elderly. Further, it is unknown if patients with extremely low levels of initial functioning simply cannot benefit from reality orientation, or if the length of treatment must be extended beyond 16 weeks.

A word of caution is necessary with respect to the widespread application of reality orientation in nursing homes and other geriatric institutions. The reality orientation procedure may undergo drastic changes as it is moved from one institution to another. The treatment originally devised at Tuscaloosa VA Hospital may not be the treatment implemented elsewhere. Based on their

observations of reality orientation at two nursing homes in the
midwest, Gubrium and Ksander (1975) noted that, for example, the
staff often required patients to respond with "right" but unrealistic answers. If the reality orientation board said it was rain-
ing, but indeed it was sunny outside, the "rewarded" response to
questions concerning the weather centered around rain. Although
Gubrium's and Ksander's example may sound a bit far-fetched, such
rigid following of established procedures is all too often a characteristic of the many "therapies" practiced in institutional settings. One of the advantages of reality orientation is that it
can be applied largely by the nonprofessional staff, usually nursing assistants. However, these staff members must be properly
trained so that they are capable of flexible judgment-making.
Periodic monitoring of the classes by professional staff would
seem to be of paramount importance. Furthermore, the establishment
of a 24-hour therapeutic milieu is often extremely difficult. Although the classes may be properly administered, the remainder of
the environment may be nontherapeutic. Finally, the treatment was
designed for moderately to severely disoriented patients. If patients are assigned to reality orientation indiscriminately, less
demented patients may resent this treatment and their hostility
may generalize to the entire staff and any other forms of therapy
offered.

REMOTIVATION

Remotivation, a group technique developed by Dorothy Hoskins
Smith for use by the nursing profession in the treatment of chronic
schizophrenics, is frequently overlooked as a psychological treatment program since it is usually not defined as "therapy." The
technique is often applied to geriatric patients with a variety
of diagnoses. It is widely used by the nursing staffs of psychiatric hospitals and nursing homes and has the advantage of requiring nonprofessional leaders, usually aides. In fact, a program
at the St. Cloud, Minnesota VA Hospital employs sixth grade children as group leaders (Thralow and Watson, 1974). However, since
it contains many of the components of other types of group therapies and seems especially pertinent to certain subgroups of geriatric patients (for example, those completing a program of reality
orientation), it is reviewed here.

The remotivation procedure is based on the hypothesis that
some aspects of the chronic mental patient's personality are
basically healthy; the goal of remotivation is to activate these
"untouched" areas. The technique usually involves one hour long
small group meetings once a week which follow a relatively structured course. First the leader endeavors to create a climate of
acceptance, usually by thanking the group as a whole for coming
and greeting them individually in a warm, friendly manner. Next,

the leader provides a <u>bridge to reality</u>. This consists of reading
a poem or article to the patients while encouraging group participation in this activity. Then the leader conducts a group discussion on a selected topic, using visual aids and other objects
related to the topic. This step is called <u>sharing the world we
live in</u>. During the discussion the leader asks group members
questions concerning the topic in order to stimulate verbal responses and interaction. From the topical discussion the leader
attempts to link the discussion to past activities of the patients
which it is hoped they will rediscover. This step is called
<u>appreciation of the work of the world</u>. Finally, the leader tries
to establish a <u>climate of appreciation</u> by expressing appreciation
and personal pleasure of the patient's attendance. Toepfer,
Bicknell, and Shaw (1974), in their comparison of the remotivation
technique to behavioral therapies, describe this final step as
one of reinforcing the patient for coming to the remotivation session. Behavior modification techniques can also be employed to
increase patient interactions within the remotivation group
(Mueller and Atlas, 1972).

Long (1962), employing an elaboration of Solomon's Four-Group
Design with a large sample, found that the experimental patients'
behavior improved on a number of dimensions after a nine week
course of weekly remotivation sessions led by college students.
He also observed that chronic patients improved more than did
acute patients and that the psychiatric aides employed on wards
where remotivation groups were instituted decreased their "cus-
todial" orientation, even though not serving directly as remotivation group leaders.

Long's study examined psychiatric patients of all ages in a
large state mental hospital. Beard and Bidus (1968) examined 111
patients exposed to remotivation sessions in a geriatric unit,
including the medically infirm, in a Veterans Administration hospital. Employing a pre-post design they observed increases in the
ward nurses' ratings of social interest on the part of all patients
and increased levels of social competence in aged chronic schizophrenic patients who had been hospitalized for an extended period
of time. These two studies speak to the applicability of the
remotivation technique to long term psychiatric patients who have
become institutionalized.

Remotivation is also used in nursing home settings. One study
compared a six months course of remotivation in a three group pre-
post design which included the parameters of sex and level of
functioning on a visual motor design reproduction test similar to
many of those used to measure "organicity" (Bowers, Anderson,
Blomeier, and Pelz, 1967). Although the study is flawed in terms
of the "blindness" of the raters, female patients were found to
demonstrate greater improvements in social functioning and in-group

behavior than did male patients. Also, those patients with the
least indication of brain syndrome did not appear to benefit from
remotivation sessions as much as did those with more severe impairment.

As was mentioned above, Thralow and Watson (1974) report the
use of elementary school children as members of the remotivation
team. They compared ratings made by the project staff of experimental patients and matched controls (N = 23 pairs) after 11 weeks
of a 16-week course of remotivation and again four weeks after the
conclusion of the treatment sessions. Twenty week results were
not encouraging with respect to those dimensions tapped by the
NOSIE-30 (Nurses Observation Scale for Inpatient Evaluation). These
dimensions included social competence, interest, neatness, irritability, psychosis, and retardation. However, the responses of experimental patients--those exposed to the twice-weekly remotivation
sessions in the company of their child-partner--on the Remotivation
Self-Evaluation Scale designed specifically for use in this study
indicated positive changes with respect to interest in travel,
interpersonal relationships and feelings about hospital life.

Bovey (1971) compared the effectiveness of six weeks of intensive remotivation therapy with that of patient-staff interactions
involving reading to the patient in a pre-post design. Both remotivation and reading to the patients served to enhance scores on
the Hospital Adjustment Scale, the Draw-A-Person test, and the
Bender-Visual-Motor Gestalt Test, in comparison to scores obtained
by a control group. Remotivation was significantly more effective
than merely reading to the patients only with respect to the measure of self-concept.

In summary, it would appear that remotivation, as currently
widely employed by the nursing staffs of psychiatric hospitals
and nursing homes, may prove useful for particular types of aged
patients with respect to particular dimensions. It should not be
seen as a panacea and deserves more detailed research in order to
determine what aspects of the procedure lead to the greatest benefits for which types of patients and which components may be
unnecessary or detrimental. Also, it is the author's observation
that the remotivation sessions are frequently seen by the staff as
just another chore they must perform as part of the work routine.
Thus, although the procedure may be quite beneficial initially
because of the interest and expectations newly initiated remotivation groups generate, it would be most useful to determine if
patient gains are also obtained when new patients are placed in
remotivation groups conducted by experienced leaders in settings
where remotivation is a routine treatment modality.

ART THERAPY

Art therapy has much in common with other forms of group therapy in that social interactions among the patients in the group may facilitate and provide support for the expression of problems, memories, and emotions. Originally developed for use with schizophrenic patients, art therapy is based on the hypothesis that perception is blocked by emotional tension. Art therapy has recently been applied to geriatric patients with both physical and mental impairment (Dewdney, 1973; Crosson, 1976; Zieger, 1976). For example, Dewdney arranges increasingly difficult "assignments" for her patients in order to reorient them to reality and reinstitute lost cognitive and psychomotor skills. Thus, art therapy may be appropriate for geriatric patients who have experienced a great deal of deterioration due to senile dementia and other organically based impairments. Zieger (1976) suggests that art therapy may also facilitate the life review process, thus broadening its applicability to the depressed, withdrawn, apathetic aged as well.

Existing reports describing the application of art therapy to older adults are limited to therapists' descriptions of the procedure and case studies. No attempt has yet been made to objectively evaluate the benefits of this promising technique for geriatric patients in nursing homes and other settings.

EXERCISE THERAPY

Increased levels of physical exercise have been found to be both feasible and beneficial to older adults in terms of physiological parameters (e.g., deVries, 1970). One study examined the influence of a program of mild physical exercise upon the cognitive and behavioral functioning of long-term geriatric patients in a state hospital (Powell, 1974). Thirty patients were assigned to one of three groups in a randomized block design: exercise therapy, social therapy, or control. The two treatment groups met for one hour a day, five days a week for a period of 12 weeks. The program of physical activity included brisk walking, calisthenics, and rhythmical movements of the arms, legs, and trunk. The social therapy program included arts and crafts, social interaction, music therapy and games. Significant improvements on two tests of cognitive function (Raven's Coloured Progressive Matrices and the Wechsler Memory Scale) were observed for the exercise therapy group in comparison to the social therapy and control groups.

Interestingly, the patients receiving exercise therapy in Powell's study obtained increasingly poorer scores on two ratings of ward behavior. Similar poor ratings of ward behavior have been noted in response to other treatments described in this chapter.

Older adults, like people of all ages, may become less passive as
they get better. They may demonstrate increased resistance to ward
routine, to established procedures and to institutionalization in
and of itself. Such acting out on the part of the institutionalized
aged may work to their disadvantage in terms of their relationships
with the institution's staff. A depressed, apathetic patient who
passively accepts the institutional regime is usually seen as a
"good" patient. He or she makes no waves. Such a patient is regarded fondly, if paternalistically. On the other hand, the dis-
satisfied, outspoken complainer who responds to treatment in a manner which, in the long run, will increase his or her level of functioning by making him more independent is all too often seen by the
staff as a "bad" or "problem" patient. Such a patient does make
waves. It takes a very enlightened staff to accept aggressive dissatisfaction as a sign of increased well-being.

Even older adults with severe physical impairments may benefit from some form of exercise. A program of mild exercise designed for nursing home residents confined to wheel chairs or
otherwise experiencing ambulation difficulties has been described
by Silverthorn, Leech, and Silverthorn (1975). The focus of the
biweekly group meetings was upon breathing exercises and the sen-
sation of minor movements. Silverthorn et al. suggest, based on
therapist observations, that such programs of sensory awareness may
improve memory and interpersonal awareness and may lead to changes
in adjustment.

RATIONAL EMOTIVE THERAPY

According to the theoretical formulations of Albert Ellis'
rational-emotive therapy, the individual experiences both rational
and irrational thoughts (subscriptions to societal myths). Internal verbalizations of the latter frequently lead to anxiety and
neurosis. The rational-emotive therapist therefore teaches the
client to think rationally.

Peth (1974) has suggested a number of age-specific myths which
may be especially troublesome to older adults. These include
views of old age as unique to a particular individual, as asexual,
as automatically producing respect and support from others, as
well as overemphasis on the importance of family ties and activity
and ideas concerning life as unchanging or a trial for which one
should be rewarded.

Only one study has attempted to evaluate the effectiveness of
rational-emotive therapy with older adults (Keller, Croake and
Brooking, 1975). Fifteen community-dwelling older adults (mean
age 68) were subjected to a four week educational program of study
and discussion of rational-emotive principles, role-playing,

reading assignments and lectures. Fifteen older adults of similar
age served as controls. The experimental program significantly
increased rational thinking, as measured by the Adult Ideas Inventory, and decreased trait anxiety. Rational-emotive therapy may
not be appropriate for those elderly experiencing significant cognitive decline; however it may serve a beneficial function for
those who are depressed, anxious and unhappy with their perceived
status in life.

FAMILY THERAPY

Although the stereotype of the older adult is one of the isolated, deserted, lonely individual, over 73 percent of the men
and 36 percent of the women who were above age 65 in 1970 were
married. Eighty percent of older adults having living siblings
and 75 percent at least one living child. Of those with children,
80 percent see at least one child on a weekly basis (Atchley,
1977). Although the classical extended family of pre-industrial
times is not prevalent, the modified extended family (Litwak,
1961) is a common occurrence in the United States. Such a family,
although not living under the same roof, may have strong emotional,
economic and support ties, as well as frequent interactions and
common activities. Berezin and Stotsky (1970) indicate that, in
fact, management of the family may be more difficult for the therapist than the treatment of the older adult patient.

Although no controlled studies of conjoint family therapy in
which the "identified" patient is an older adult have been reported, the number of case and program descriptions is growing
(e.g., Peterson, 1973). Grauer, Betts and Birnbom (1973) have
described a two-phase program of family therapy in connection
with a day hospital for older adults. Phase one involved complete or partial separation of the family members in order to reduce open conflict and hostilities. In some instances lack of
sufficient family strengths or previous good family adjustment
limited treatment to separation alone. However, in a number of
cases, a working-through phase followed in which the older adult
and members of the family were able to deal with conflicts and
needs and ultimately achieve improved communication and closeness.
Families capable of phase two treatment were described as those
with a history of welfare emotions, i.e., gratifying familial relationship including sincerity, empathy, generosity and trust.

Goldstein and Birnbom (1976) have reported good success with
family therapy wherein the "identified" older patient was suffering from hypochondriasis--a psychopathology relatively common to
older adults and for which satisfactory treatment is generally
lacking. Intergenerational family therapy may also prove useful
in instances where the family dynamics intimately involve the

older generation, although the "identified" patient may be a member of the second or third generation (Spark, 1973).

Family attitudes toward the older family member, and hence the potential benefits of family therapy, may depend upon the identified patient's diagnostic classification. For example, Baer, Morin and Gaitz (1970) found that family attitude was more positive with respect to older patients suffering from organic brain syndrome than with respect to those older adults suffering from alcoholism or functional psychosis.

HYPERBARIC OXYGENATION

A treatment approach quite distinct from the others described in this chapter involves the administration of 100% oxygen at greater than normal atmospheric pressure to elderly individuals demonstrating poor cognitive functioning. The rationale for the treatment involves the hypothesis that cognitive dysfunction in some older adults, particularly those diagnosed as suffering from chronic brain syndromes, may be due to vascular disease and that the function of remaining brain cells can be improved if the needed oxygen is supplied. Improved brain cell function would then be reflected in improved cognitive function.

The first report of this treatment procedure indicated improvements on several cognitive measures in treated patients as compared to untreated controls (Jacobs, Winter, Alvis, and Small, 1969). Subsequent well-controlled studies, however, have not replicated these results (Goldfarb, Hochstadt, Jacobson, and Weinstein, 1972; Thompson, 1975). The major research issues concerning the evaluation of the benefits of this treatment have centered around the appropriate use of the double-blind procedure and of controls. Further, it may well be that an examination of the etiology of the cognitive dysfunction is of primary importance. Some types of chronic brain syndrome in old age appear to stem from degenerative, as opposed to vascular, changes in the brain. Increased oxygenation of brain cells in such cases would be ineffective.

SUMMARY

It can be seen from the review of the "other approaches" included in this chapter that numerous efforts are now being directed towards developing appropriate, innovative ways of meeting the mental health needs of older adults, primarily the more obvious ones of the institutionalized aged. However, this effort has just begun; the research with respect to these treatments is largely preliminary and, with some exceptions, relatively unso-

phisticated. Clinical psychology has long ignored the treatment
of older adults. The gap has frequently been filled by those from
other disciplines who have recognized the need but are less attuned to the requirements of experimental design and research
methodology in clinical settings. Thus, it is hoped that this
chapter will serve two purposes: First, that it will point the
service provider to the various treatment options which have been
suggested for the geriatric patient. Second, that it will stimulate clinical research psychologists to roll up their sleeves and
get to work.

REFERENCES

Atchley, R. C. The social forces in later life (2nd ed.). Bel-
    mont, Calif.: Wadsworth Publishing Co., 1977.
Baer, P. E., Morin, K., and Gaitz, C. M. Familial resources of
    elderly psychiatric patients. Archives of General Psychi-
    atry, 1970, 22, 343-350.
Barnes, J. A. Effects of reality orientation classroom on memory
    loss, confusion, and disorientation in geriatric patients.
    The Gerontologist, 1974, 14, 138-142.
Beard, M. T., and Bidus, D. R. A study of the effects of remoti-
    vation on social competence, social interest and personal
    neatness. Journal of Psychiatric Nursing and Mental Health
    Services, 1968, 6, 197-201.
Berezin, M. A., and Stotsky, B. A. The geriatric patient. In H.
    Grunebaum (Ed.), The practice of community mental health.
    Boston: Little, Brown and Co., 1970.
Bok, M. Some problems in milieu treatment of the chronic older
    mental patient. The Gerontologist, 1971, 11, 141-147.
Bovey, J. A. The effect of intensive remotivation techniques on
    institutionalized geriatric mental patients in a state mental
    hospital. Unpublished doctoral dissertation, North Texas
    State University, 1971.
Bowers, M. B., Anderson, G. K., Blomeier, E. C., and Pelz, K.
    Brain syndrome and behavior in geriatric remotivation groups.
    Journal of Gerontology, 1967, 22, 348-352.
Brook, P., Degun, G., and Mather, M. Reality orientation, a ther-
    apy for psychogeriatric patients: A controlled study.
    British Journal of Psychiatry, 1975, 127, 42-45.
Citrin, R. S., and Dixon, D. N. Reality orientation: A milieu
    therapy used in an institution for the aged. The Gerontolo-
    gist, 1977, 17, 39-43.
Crosson, C. Art therapy with geriatric patients: Problems of
    spontaneity. American Journal of Art Therapy, 1976, 15,
    51-56.
deVries, H. A. Physiological effects of an exercise training
    regimen upon men 52-88. Journal of Gerontology, 1970, 25,
    325-336.

Dewdney, I. An art therapy program for geriatric patients. American Journal of Art Therapy, 1973, 12, 249-254.

Folsom, J. C. Intensive hospital therapy of geriatric patients. In J. H. Masserman (Ed.), Current psychiatric therapies (Vol. 7). New York: Grune & Stratton, 1967.

Gatz, M., Siegler, I. C., and Dibner, S. S. Individual and community: Normative conflicts in the development of a new therapeutic community. Manuscript submitted for publication, 1977.

Goldfarb, A. I., Hochstadt, N. J., Jacobson, J. H., and Weinstein, E. A. Hyperbaric oxygen treatment of organic mental syndrome in aged persons. Journal of Gerontology, 1972, 27, 212-217.

Goldstein, S. W., and Birnbom, F. Hypochondriasis and the elderly. Journal of the American Geriatrics Society, 1976, 24, 150-154.

Gottesman, L. E. Resocialization of the geriatric mental patient. American Journal of Public Health, 1965, 55, 1964-1970.

Gottesman, L. E. The response of long-hospitalized aged psychiatric patients to milieu treatment. The Gerontologist, 1967, 7, 47-48.

Grauer, H., Betts, D., and Birnbom, F. Welfare emotions and family therapy in geriatrics. Journal of the American Geriatrics Society, 1973, 21, 21-24.

Gubrium, J. F., and Ksander, M. On multiple realities and reality orientation. The Gerontologist, 1975, 15, 142-145.

Gutmann, D., Gottesman, L., and Tessler, S. A comparative study of ego functioning in geriatric patients. The Gerontologist, 1973, 13, 419-423.

Harris, C. S., and Ivory, P. B. C. B. An outcome evaluation of reality orientation therapy with geriatric patients in a state mental hospital. The Gerontologist, 1976, 16, 496-503.

Jacobs, E. A., Winter, P. M., Alvis, H. J., and Small, S. M. Hyperoxygentation effect on cognitive functioning in the aged. New England Journal of Medicine, 1969, 281, 753-757.

Keller, J. F., Croake, J. W., and Brooking, J. Y. Effects of a program in rational thinking on anxieties in older persons. Journal of Counseling Psychology, 1975, 22, 54-57.

Liederman, P. C., and Liederman, V. R. Group therapy: An approach to problems of geriatric outpatients. In J. H. Masserman (Ed.), Current psychiatric therapies (Vol. 7). New York: Grune & Stratton, 1967.

Litwak, E. Geographical mobility and family cohesion. American Sociological Review, 1961, 26, 258-271.

Long, R. Remotivation--Fact or artifact. Washington, D.C.: American Psychiatric Association, Supplementary Mailing No. 151, 1962.

Lowenthal, M. F., and Haven, C. Interaction and adaptation: Intimacy as a critical variable. American Sociological Review, 1968, 33, 20-30.

Mishara, B. L. Geriatric patients who improve in token economy and general milieu treatment programs: A multivariate analysis. Paper presented at the meeting of the American Psychological Association, San Francisco, August, 1977.

Mueller, D. J., and Atlas, L. Resocialization of regressed elderly residents: A behavioral management approach. Journal of Gerontology, 1972, 27, 390–392.

Oberleder, M. Crisis therapy in mental breakdown of the aging. The Gerontologist, 1970, 10, 111–114.

Peterson, J. A. Marital and family therapy involving the aged. The Gerontologist, 1973, 13, 27–31.

Peth, P. R. Rational-emotive therapy and the older adult. Journal of Contemporary Psychotherapy, 1974, 6, 179–184.

Powell, R. R. Psychological effects of exercise therapy upon institutionalized geriatric mental patients. Journal of Gerontology, 1974, 29, 157–161.

Reichenfeld, H. F., Csapo, K. G., Carriere, L., and Gardner, R. C. Evaluating the effect of activity programs on a geriatric ward. The Gerontologist, 1973, 13, 305–310.

Sanders, R., Smith, R. S., and Weinman, W. S. Chronic psychosis and recovery. San Francisco: Jossey-Bass, 1967.

Silverthorn, A. I., Leech, S., and Silverthorn, L. J. Use of sensory awareness with geriatric patients. Paper presented at the meeting of the American Psychological Association, Chicago, August, 1975.

Smyer, M. A., Siegler, I. C., and Gatz, M. Learning to live in a therapeutic community: A study of elderly inpatients. International Journal of Aging and Human Development, 1976, 7, 231–235.

Spark, G. M. Grandparents and intergenerational family therapy. Paper presented at the annual convention of the American Psychological Association, Montreal, 1973.

Spence, D. L., Cohen, S., and Kowalski, C. Mental health, age, and community living. The Gerontologist, 1975, 15, 77–82.

Steer, R. A., and Boger, W. P. Milieu therapy with psychiatric-medically infirm patients. The Gerontologist, 1975, 15, 138–141.

Taulbee, L. R., and Folsom, J. C. Reality orientation for geriatric patients. Hospital and Community Psychiatry, 1966, 17, 133–135.

Taulbee, L. R., and Wright, H. W. A psychosocial-behavioral model for therapy. In C. W. Spielberger, Current topics in clinical and community psychology (Vol. 3). New York: Academic Press, 1971.

Thompson, L. W. Effects of hyperbaric oxygen on behavioral functioning in elderly persons with intellectual impairment. In S. Gershon and A. Raskin (Eds.), Aging, Vol. 2. New York: Raven Press, 1975.

Thralow, J. U., and Watson, C. G. Remotivation for geriatric patients using elementary school students. The American Journal of Occupational Therapy, 1974, 28, 469–473.

Toepfer, C. T., Bicknell, A. T., and Shaw, D. O. Remotivation as behavior therapy. The Gerontologist, 1974, 14, 451–453.

Zieger, B. L. Life review in art therapy with the aged. American Journal of Art Therapy, 1976, 15, 47–50.

LIST OF CONTRIBUTORS

Donna Cohen, Ph.D., Assistant Professor of Psychiatry and Behavioral Sciences, and Chief, Behavioral Biology Unit, Geriatric Research Educational and Clinical Center, Seattle/American Lake Veterans Administration Hospitals, Department of Psychiatry and Behavioral Sciences, School of Medicine, University of Washington, Seattle, Washington.

Paul T. Costa, Jr., Ph.D., Associate Professor of Psychology, Psychology Department, University of Massachusetts-Boston, Boston, Massachusetts.

Carl Eisdorfer, Ph.D., M.D., Professor and Chairman, Department of Psychiatry and Behavioral Sciences, School of Medicine, University of Washington, Seattle, Washington.

Merrill F. Elias, Ph.D., Professor of Psychology, Department of Psychology, University of Maine at Orono, Orono, Maine.

W. Doyle Gentry, Ph.D., Professor of Medical Psychology, Department of Psychiatry, and Senior Fellow in the Center for the Study of Aging and Human Development, Duke University, Durham, North Carolina.

Boaz Kahana, Ph.D., Professor of Psychology, Oakland University, Rochester, Michigan.

Robert L. Kahn, Ph.D., Associate Professor, Departments of Psychiatry and Behavioral Science (Human Development), University of Chicago, Chicago, Illinois.

Philip L. Kapnick, Ph.D., Associate Professor, St. Louis College of Pharmacy, St. Louis, Missouri.

Robert Kastenbaum, Ph.D., Superintendent, Cushing Hospital, Framingham, Mass.; Professor of Psychology, University of Massachusetts-Boston, Boston, Massachusetts.

Diane K. Klisz, Ph.D., Gerontological Psychologist, Geriatric Research, Education and Clinical Center, Veterans Administration Hospital, St. Louis, Missouri.

Robert R. McCrae, Ph.D., Project Director, Smoking and Personality Research Grant, University of Massachusetts-Boston, Boston, Massachusetts.

Nancy E. Miller, Ph.D., Head, Section on Clinical Research, Center for Studies of the Mental Health of the Aging, National Institute of Mental Health, Rockville, Maryland.

William S. Richards, Ph.D., Director, The Elderly Services Program, The Counseling Center, Bangor, Maine.

Ilene C. Siegler, Ph.D., Assistant Professor of Medical Psychology, Duke University Medical Center, and Senior Fellow, Center for the Study of Aging and Human Development, Duke University, Durham, North Carolina.

Martha Storandt, Ph.D., Associate Professor of Psychology, Aging and Development Program, Department of Psychology, Washington University, St. Louis, Missouri.

Geoffrey L. Thorpe, Ph.D., Clinical Director of Outpatient Services, The Counseling Center, Bangor, Maine, and Cooperating Assistant Professor of Psychology, Department of Psychology, University of Maine at Orono, Orono, Maine.

W. Gibson Wood, Ph.D., Postdoctoral Fellow, National Institute on Alcohol Abuse and Alcoholism, and Cooperating Assistant Professor of Psychology, Department of Psychology, University of Maine, Orono, Maine.